STUDY GUIDE

for

COLLEGE ALGEBRA
Kolman/Shapiro

by

James W. Snow

Whatcom Community College

Academic Press

A Subsidiary of Harcourt Brace Jovanovich, Publishers

New York London Toronto Sydney San Francisco

Academic Press, Inc.
111 Fifth Avenue, New York, New York 10003

United Kingdom Edition published by
Academic Press, Inc. (London) Ltd.
24/28 Oval Road, London NW1

ISBN: 0-12-417887-1

Manufactured in the United States of America.

PREFACE

In an attempt to assist you in your learning of mathematics, this study guide is written as a supplement to the basic text, not a replacement for it. You will find it easy to move back and forth between the text and the study guide since the language is very similar and the chapter and section numbers correspond.

In a very real sense this study guide is more of a solutions manual than it is a step-by-step guide through the material, since a large portion of the book is devoted to detailed solutions to the exercises found in the text. Did you ever just want to peek at an answer or a solution to see what it was, or to get a hint how to start? Probably so, but if you are like many students you thought that was cheating or not fair, or at least not a good thing to do. Right? Well, in my opinion, it is not only all right to do, but it is an excellent way to learn a lot of mathematics. After all, what do you do in math class when you don't understand the problem? You ask the professor to do it on the board, of course. And what does he then do? Most likely he will work it out to your satisfaction. So, indeed, I urge you to use the solutions as a learning device to help you see a way to approach a problem.

You will find the Study Guide and Solutions Manual consists of four major components:

1. The BASIC CONCEPTS you should learn from each unit;

2. WHAT YOU SHOULD KNOW HOW TO DO upon completion of each unit;

3. SOLUTIONS to selected problems from the text; and

4. A short CHAPTER QUIZ covering concepts and problem types. Answers to the quizzes are included.

There are a number of uses for a study guide and solutions manual such as this one; among these are:

1. Use as a learning tool as you proceed through the course;

2. Use as a learning tool to review just before a chapter test, a mid-term exam, and before the final examination;

3. Use as a reference book if you have difficulty understanding a concept or in solving a problem; and

4. Use as a quick overview before you begin the study of a section in the text.

Somewhere in this work you may find an error or two. They are probably mine, and I apologize for them in advance. I know all too well how very frustrating it is to be working along and then find out that 'the book is wrong.'

It is my sincere hope that by having this study guide and solutions manual available your learning of college algebra and trigonometry will be a bit easier and more enjoyable.

Acknowledgements: I wish to thank especially the editorial staff of Academic Press, Inc. who offered many constructive comments on the preparation of the manuscript, and also to Ms. Margaret Fogg who did a great job of typing the material under a rather tight time schedule.

Jim Snow

Bellingham, Washington

February, 1981

Table of Contents

CHAPTER 1

The Foundations of Algebra

Section 1.1 The Real Number System

BASIC CONCEPTS you should learn from this unit.

1. Symbols and vocabulary of sets

2. The fundamental subsets of the real number system are Natural Numbers, Integers, Rational Numbers, Irrational Numbers.

3. The twelve axioms of a field

4. The four properties of equality

WHAT YOU SHOULD KNOW HOW TO DO upon completion of this unit.

1. Use set notation

2. Identify various numbers as being real, integers, rational, irrational

3. Identify the properties of real numbers

4. Identify and use the properties of equality.

SOLUTIONS to exercise set 1.1

2. Integers between -4 and 2 would be $\{-3, -2, -1, 0, 1\}$.

5. Only the positive integers of $S = \{1,2\}$.

8. Since x must be even, we have $\{2,8,10\}$.

11. F π is irrational, so $\pi/2$ is also irrational.

14. F $0.75 = \frac{75}{100}$, so is rational.

17. T $\sqrt{5}$ is an irrational and all irrationals and rationals are real.

20. T

23. Commutative property of addition

26. This is the closure property of addition of two real numbers.

29. Closure for multiplication

32. Zero is the additive identity element.

35. Multiplicative inverse property (property 10)

38. For example, let $a = 1$, $b = 3$; $\frac{1}{3} \neq \frac{3}{1}$.

41. Symmetric property: $a = b$, then $b = a$.

44. Substitution property

47. Given $a + c = b + c$
then $a + c + (-c) = b + c + (-c)$ — Theorem in text
and $a + 0 = b + 0$ — Additive inverse
thus $a = b$ — Additive identity

50. Using the hint, we have $b = \frac{1}{0}$ is the reciprocal of 0.

$1 = 0 \cdot \frac{1}{0}$ — Property 10 A number times its reciprocal = 1.

$1 = 0 \cdot b$ — Substitution

$1 = 0$ — See Theorem 3, (the right side will equal 0 because one factor equals 0.)

Section 1.2 Real Number Line

BASIC CONCEPTS you should learn from this unit:

1. Representation of real numbers geometrically on a number line
2. Properties of inequality, especially property (e): If $a < b$, then $ac > bc$ if $c < 0$. (Multiplying by a negative reverses the inequality.)
3. Concept of absolute value, and properties.

WHAT YOU SHOULD KNOW HOW TO DO upon completion of this unit:

1. Locate real numbers on a number line
2. Designate the relative size of two numbers using inequality symbols, as $-3 < 0.2$
3. Indicate sets of real numbers on a number line
4. Translate word statements into equalities and vice versa
5. Determine the absolute value of a given number
6. Find the distance between two points on a number line using the absolute value.

SOLUTIONS to Exercise Set 1.2:

2.

5. 4 would be to the left of 6.

8. -4 would be to the left of 0.

11.
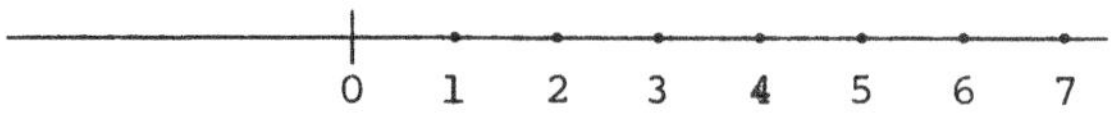

Section 1.2 Real Number Line

14. 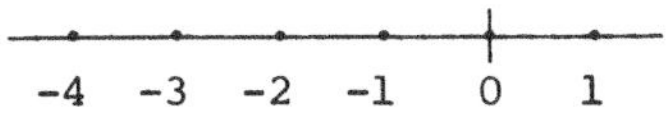

17. $a \geq 0$; that is, a is greater than or equal to zero.

20. $3 < a < 7$

23. $b \geq 5$.

26. Transitive property

29. Property (d); multiplying by a number greater than zero.

32. 2/3

35. -2

38. $|0| = 0$.

41. $\frac{|6|}{|-3|} = \frac{6}{3} = 2$.

44. $\frac{4 - 2 \cdot 4 \cdot 3}{1} = 4 - 24 = -20$.

47. $|-1 - (-3)| = |2| = 2$.

Section 1.3 Polynomials

BASIC CONCEPTS you should learn from this unit.

1. Definition of algebraic expression: variables and constants with the operations of +, −, x, ÷
2. Evaluation of an expression
3. Definition of an exponent
4. Definition of monomial and of polynomial as the sum of monomials
5. Degree of a monomial and polynomial
6. Properties of the real numbers applied to polynomials
7. Operations of addition and multiplication with polynomials.

WHAT YOU SHOULD KNOW HOW TO DO upon completion of this unit.

1. Define "Algebraic Expression."
2. Find the numeric value of an expression given the value(s) of the variable(s).
3. Recognize monomials and polynomials.
4. Add and multiply polynomials.

SOLUTIONS to Exercise Set 1.3:

2. $rst = 2 \cdot 3 \cdot 4 = 24$.

5. $\frac{r + s}{rt} = \frac{2 + 3}{2 \cdot 4} = \frac{5}{8}$.

8. $c = 37$, so we have $\frac{9}{5} \cdot 37 + 32 = \frac{333}{5} + 32 = \underline{98.6}$.

11. Substituting yields $(0.02)(2.5) + (0.314)(3.4)(2.81) + (2.25)(2.81)$

$$= 0.05 + 2.999956 + 6.3225$$

$$= 9.372456.$$

14. $|-3 + 2| + |-3 -2| = |-1| + |-5| = 6$.

17. $\frac{-|-2 -2|}{|-2 -1|} = \frac{-4}{3}$.

20. $x^3 \cdot x^5 = x^{3 + 5} = x^8$.

23. $\left(\frac{3}{2} x^3\right)\left(-2x\right) = -3x^4$.

26. (a) and (d) are not polynomials because of fractional exponents.

29. Leading coefficient is $\frac{3}{5}$. Degree is 4.

32. Degree is 4 (from the first term $4xy^3 = 4x^1y^3$).

35. Substituting, we obtain $3 \cdot 2^2 \cdot (-1)^2 + 2 \cdot 2(-1) - 2 + 2(-1) + 7$

$$= 12 - 4 - 2 - 2 + 7 = 11.$$

38. πr^2

41. The expression $55x + 45y + 60z$ represents the total cost of all stocks bought.

44. Remove parentheses: $4xy^2 + 2xy + 2x + 3 + 2xy^2 - xy + y - 2$; then combine terms to get $6xy^2 + xy + 2x + y + 1$.

47. Combine like terms: $-2a^2bc + ab^2c - 2ab^3 + 3$.

50. $(2s - 3)(s^3 - s + 2) = 2s^4 - 2s^2 + 4s - 3s^3 + 3s - 6 = 2s^4 - 3s^3 - 2s^2 + 7s - 6$.

53. $(2y^2 + y)(- 2y^3 + y - 3) = - 4y^5 + 2y^3 - 6y^2 - 2y^4 + y^2 - 3y$

$$= - 4y^5 - 2y^4 + 2y^3 - 5y^2 - 3y.$$

56. $(2a^2 + ab + b^2)(3a - b^2 + 1) =$
$6a^3 - 2a^2b^2 + 2a^2 + 3a^2b - ab^3 + ab + 3ab^2 - b^4 + b^2$. There are no like terms to combine.

59. $2(3x - 2)(2x + 3) = (6x - 4)(2x + 3) = 12x^2 + 18x - 8x - 12 =$
$12x^2 + 10x - 12.$

62. A = total area - area of cutout; $A = xy - \left(\frac{x}{2}\right)^2.$

65. Mentally you would think $4x^2 + 6x + 2x + 3 = 4x^2 + 8x + 3.$

68. $(x + 4)(2x - 1) = 2x^2 - x + 8x - 4 = 2x^2 + 7x - 4.$

71. $(3x - 1)(3x - 1) = 9x^2 - 3x - 3x + 1 = 9x^2 - 6x + 1.$

74. $(3a + 2b)(3a + 2b) = 9a^2 + 6ab + 6ab + 4b^2 = 9a^2 + 12ab + 4b^2.$

Section 1.4 Factoring

BASIC CONCEPTS you should learn from this unit.

1. Factoring is a process of finding the factors of an expression.
2. Factoring a common factor
3. Factoring trinomials
4. Factoring the difference of two squares: $a^2 - b^2 = (a + b)(a - b).$

WHAT YOU SHOULD KNOW HOW TO DO upon completion of this unit.

1. Define factoring.
2. Factor polynomials of the types in this unit, that is, common factor, difference of two squares, trinomials.

SOLUTIONS to Exercise Set 1.4.

2. $\frac{1}{2}x + \frac{1}{4}y = \frac{1}{2}\left(x + \frac{1}{2}y\right)$ Common factor

5. $5bc + 25b = 5b(c + 5)$ Common factor

8. $3abc + 12bc = 3bc(a + 4).$

11. $x^2 + 4x + 3 = (x + \quad)(x + \quad)$
$= (x + 3)(x + 1).$

14. $y^2 + 7y - 8 = (y + \quad)(y - \quad)$
$= (y + 8)(y - 1).$

17. $y^2 - \frac{1}{9} = \left(y + \frac{1}{3}\right)\left(y - \frac{1}{3}\right)$ Difference of two squares.

20. $4b^2 - a^2 = (2b + a)(2b - a)$ Difference of two squares

23. $\frac{1}{16} - y^2 = (\frac{1}{4} + y)(\frac{1}{4} - y)$ Like Exercise 17

26. $a^2b^2 - \frac{1}{9} = (ab + \frac{1}{3})(ab - \frac{1}{3})$.

29. $x^2 + 11x + 24 = (x + \quad)(x + \quad)$

$= (x + 8)(x + 3)$.

32. $2x^2 + 7x + 6 = (2x \quad)(x \quad)$

$= (2x + 3)(x + 2)$.

35. $6x^2 + 13x + 6 = (3x + 2)(2x + 3)$

38. $9x^2 + 24x + 16 = (3x + 4)^2$ A square trinomial

41. $6a^2 - 5ab - 6b^2 \quad (3a + 2b)(2a - 3b)$ (Hint: See Exercise #35).

44. $x^{12} - 1 = (x^6 - 1)(x^6 + 1)$ Difference of two squares. $(x^6 - 1)$ also is difference of two squares.

$= (x^3 + 1)(x^3 - 1)(x^6 + 1)$.

47. $8n^2 - 18n - 5 = (2n - 5)(4n + 1)$

50. $3y^2 + 6y - 45 = 3(y^2 + 2y - 15)$ Common factor first

$= 3(y + 5)(y - 3)$

53. $18x^2m + 33xm + 9m = 3m(6x^2 + 11x + 3)$

$= 3m(3x + 1)(2x + 3)$

56. $10r^2 - 5rs - 15s^2 = 5(2r^2 - rs - 3s^2)$

$= 5(2r - 3s)(r + s)$.

59. $b^4 + 2b^2 - 8 = (b^2 \quad)(b^2 \quad)$

$= (b^2 + 4)(b^2 - 2)$.

61. $6b^4 + 7b^2 - 3 = (3b^2 - 1)(2b^2 + 3)$.

64. $3(x + 2)^2(x - 1) - 4(x + 2)^2(2x + 7) = (x + 2)^2[3(x - 1) - 4(2x + 7)]$

$= (x + 2)^2(-5x - 31)$

$= -(x + 2)^2(5x + 31)$

BASIC CONCEPTS you should learn from this unit.

1. Definition of a rational expression; that is, the quotient of two polynomials
2. Multiplication and division are carried out much the same as is done with arithmetic fractions.
3. Addition and subtraction, too, are handled as in arithmetic, except that more care is needed in finding and using the lowest common denominator (LCD).
4. Remember that polynomials and thus rational expressions merely represent real numbers, so the same properties will apply.

WHAT YOU SHOULD KNOW HOW TO DO upon completion of this unit.

1. Multiply, divide, add, and subtract rational algebraic expressions.
2. Given two or more rational expressions, find the lowest common denominator.
3. Simplify complex fractions.

SOLUTIONS to Exercise Set 1.5

2. $\frac{y^2 - 25}{y + 5} = \frac{\cancel{(y + 5)}(y - 5)}{\cancel{(y + 5)}} = y - 5.$

5. $\frac{6x^2 - x - 1}{2x^2 + 3x - 2} = \frac{(3x + 1)\cancel{(2x - 1)}}{\cancel{(2x - 1)}(x + 2)} = \frac{3x + 1}{x + 2}.$

8. Invert divisor: $\frac{5\cancel{(x + 3)}}{8_2} \cdot \frac{4}{3\cancel{(x + 3)}} = 5/6.$

11. Invert divisor: $\frac{\cancel{(x + 2)}}{\cancel{3y}} \cdot \frac{\cancel{3y} \cdot 5y}{\cancel{(x + 2)}(x - 4)} = \frac{5y}{x - 4}.$

14. Factor all parts: $\frac{6x^2 + 7x - 2}{(4x + 1)\cancel{(x - 1)}} \cdot \frac{(5x + 2)\cancel{(x - 1)}}{(3x + 1)(x + 2)} = \frac{(6x^2 + 7 - 2)(5x + 2)}{(4x + 1)(3x + 1)(x + 2)}.$

17. Invert divisor: $\frac{\cancel{(x - 5)}(x + 3)}{1} \cdot \frac{(x^2 + 1)}{\cancel{(x - 5)}(x - 2)} = \frac{(x + 3)(x^2 + 1)}{(x - 2)}.$

20. Factor: $\frac{\cancel{(x + 3)}\cancel{(x - 3)}}{(3x - 1)\cancel{(2x + 1)}} \cdot \frac{\cancel{(2x + 1)}(x + 2)}{\cancel{(x + 3)}\cancel{(x + 1)}} \cdot \frac{(x - 2)\cancel{(x + 1)}}{x\cancel{(x - 3)}} = \frac{x^2 - 4}{3x^2 - x}.$

23. LCD = $\underline{2a}$

26. $(x^2 - 4) = (x + 2)(x - 2)$
$(x - 2) = (x - 2)$
Therefore, LCD = $(x + 2)(x - 2)$.

29. All denominators are needed: LCD = $x(x + 1)(x - 1)$.

32. $\frac{x}{x^2 - 4} + \frac{2}{4 - x^2} = \frac{x}{x^2 - 4} - \frac{2}{x^2 - 4} = \frac{(x - 2)}{(x + 2)(x - 2)} = \frac{1}{x + 2}$

35. $\frac{1}{a + 2} + \frac{3}{a - 2}$ The LCD = $(a + 2)(a - 2)$, so we have

$\frac{a - 2}{LCD} + \frac{3(a + 2)}{LCD} = \frac{a - 2 + 3a + 6}{LCD} = \frac{4a + 4}{(a + 2)(a - 2)}$

38. $\frac{4x - 1}{6x^3} + \frac{2}{3x^2} = \frac{4x - 1}{6x^3} + \frac{4x^4}{6x^3} = \frac{8x - 1}{6x^3}$ The LCD = $6x^3$

41. $\frac{5x}{2(x + 3)(x - 3)} + \frac{4}{3(x^2 + 3)}$ The LCD = $2 \cdot 3(x + 3)(x - 3)(x^2 + 3)$, so we have

$\frac{5x(3)(x^2 + 3) + 4(2)(x + 3)(x - 3)}{LCD} = \frac{15x^3 + 45x + 8x^2 - 72}{6(x + 3)(x - 3)(x^2 + 3)}$

44. $\frac{2x}{(2x + 1)} - \frac{x-1}{(2x + 1)(x - 2)}$ The LCD = $(2x + 1)(x - 2)$, so we have

$\frac{2x(x - 2) - (x - 1)}{LCD} = \frac{2x^2 - 5x + 1^{12}}{(2x + 1)(x - 2)}$.

47. $\frac{2x - 1}{(x + 3)(x + 2)} - \frac{x - 2}{(x + 3)(x + 1)}$ The LCD = $(x + 3)(x + 2)(x + 1)$, so we have

$\frac{(2x - 1)(x + 1) - (x - 2)(x + 2)}{LCD} = \frac{2x^2 + x - 1 - x^2 + 4}{LCD} = \frac{x^2 + x + 3}{(x + 3)(x + 2)(x + 1)}$

50. In this problem the LCD = $(x + 2)(x - 2)$, so we have

$\frac{2x(x - 2) + x(x + 2) - 1}{LCD} = \frac{3x^2 - 2x - 1}{x^2 - 4}$.

53. $\frac{(x + 1)}{(1 - \frac{1}{x})} \cdot \frac{x}{x} = \frac{x^2 + x}{x - 1}$.

56. $\frac{\frac{a}{a - b} - \frac{b}{a + b}}{a^2 - b^2} = \frac{a^2 + ab - ab + b^2}{a^2 - b^2} \cdot \frac{1}{a^2 - b^2} = \frac{a^2 + b^2}{(a^2 - b^2)^2}$.

59. $\frac{(\frac{a}{b} - \frac{b}{a})}{(\frac{1}{a} + \frac{1}{b})} \cdot \frac{ab}{ab} = \frac{a^2 - b^2}{b + a} = \frac{\cancel{(a + b)}(a - b)}{\cancel{(a + b)}} = a - b.$

62. $2 + \dfrac{3}{1 + \dfrac{2}{1 - x}} = 2 + \dfrac{3(1 - x)}{1 - x + 2} = 2 + \dfrac{3 - 3x}{3 - x}$

$= \dfrac{2(3 - x)}{3 - x} + \dfrac{3 - 3x}{3 - x} = \dfrac{9 - 5x}{3 - x}.$

65. $1 - \dfrac{1}{1 + \dfrac{1}{1 - \dfrac{1}{1 + x}}} = 1 - \dfrac{1}{1 + \dfrac{1 + x}{1 + x - 1}} = 1 - \dfrac{1}{1 + \dfrac{1 + x}{x}} = 1 - \dfrac{x}{x + 1 + x}$

$= 1 - \dfrac{x}{1 + 2x} = \dfrac{1 + 2x - x}{1 + 2x} = \dfrac{1 + x}{1 + 2x}$

Section 1.6 Integer Exponents

BASIC CONCEPTS you should learn from this unit.

1. The rules for positive integer exponents.
2. The meaning of the zero exponent; that is, $x^0 = 1$.
3. The meaning and use of negative integer exponents; for example, $x^{-2} = \dfrac{1}{x^2}$.

WHAT YOU SHOULD KNOW HOW TO DO upon completion of this unit.

1. Simplify expressions containing integer exponents.
2. Change an expression with negative exponents to one containing only positive exponents.

SOLUTIONS to Exercise Set 1.6

2. Multiply 2 times 5 to obtain the correct exponent

   Answer = y^{10}

5. $(2x)^4 = 2^4x^4 = 16x^4$.

8. x^{3m^2}

11. $-\dfrac{x^3}{y^3}$

14. $x^{12-8} = x^4$.

17. $x^{3m + m} = x^{4m}$.

20. $\left(\dfrac{3x^3}{y^2}\right)^5 = \dfrac{3^5x^{15}}{y^{10}}$

23. $\dfrac{(r^2)^4}{(r^4)^2} = \dfrac{r^8}{r^8} = 1.$

26. $\dfrac{(-2a^2b)^4}{(-3ab^2)^3} = \dfrac{16a^8b^4}{-27a^3b^6} = \dfrac{16a^5}{-27b^2}$

29. $(-2a^2b^3)^{2m} = (-2)^{2m}a^{4m}b^{6m}.$

32. $(xy)^0 - 2^{-1} = 1 - \frac{1}{2} = \frac{1}{2}.$

35. $\dfrac{1}{3^{-4}} = 3^4 = 81.$

38. $-x^{-5} = -1/x^5.$

41. $5^{-3} \cdot 5^5 = 5^{-3+5} = 5^2 = 25.$

44. $(x^{-2})^4 = x^{-8} = 1/x^8.$

47. $\dfrac{2^2}{2^{-3}} = 2^{2-(-3)} = 2^5 = 32.$

50. $(x^4y^{-2})^{-1} = x^{-4}y^2 = \dfrac{y^2}{x^4}.$

53. $\left(-\frac{1}{2}x^3y^{-4}\right)^{-3} = \left(-\frac{1}{2}\right)^{-3}x^{-9}y^{12} = \dfrac{-8y^{12}}{x^9}.$

56. $\left(\dfrac{x^3}{x^{-2}}\right)^2 = \left(x^5\right)^2 = x^{10}.$

59. $(a - 2b^2)^{-1} = \dfrac{1}{a - 2b^2}$

62. $(a^{-1} + b^{-1})^{-1} = \dfrac{1}{a^{-1} + b^{-1}} = \dfrac{1}{\frac{1}{a} + \frac{1}{b}} = \dfrac{1}{\frac{b+a}{ab}} = \dfrac{ab}{a+b}.$

65. $\left(\dfrac{a}{b}\right)^{-m} = \dfrac{1}{\left(\frac{a}{b}\right)^m} = \dfrac{1}{\frac{a^m}{b^m}} = \dfrac{b^m}{a^m} = \left(\dfrac{b}{a}\right)^m$ Q.E.D.

68. $\left(\dfrac{7.65^{-1}}{7.65^2}\right)^2 = \left(\dfrac{1}{7.65^3}\right)^2 = \dfrac{1}{7.65^6} = 4.99 \times 10^{-6}.$

Section 1.7 Rational Exponents and Radicals

BASIC CONCEPTS you should learn from this unit.

1. Rational exponents and roots: $b^{1/n}$ is an $n^{\underline{th}}$ root of b; for example, $b^{1/3}$ is a cube root of b.
2. Meaning of $a^{m/n}$ as $(a^m)^{1/n} = (a^{1/n})^m$
3. Since $\sqrt{b} \cdot \sqrt{b} = b$ and since $b^{1/2} \cdot b^{1/2} = b$, $\sqrt{b} = b^{1/2}$. In general, $\sqrt[n]{b} = b^{1/n}$.
4. Meaning of a simplified radical.
5. Radicals can be written in exponent form, so that the rules of exponents can be used in simplifying.
6. Meaning of "Rationalize the denominator."

WHAT YOU SHOULD KNOW HOW TO DO upon completion of this unit.

1. Simplify expressions with rational exponents
2. Change an expression with rational exponents to one with radicals and vice versa
3. Simplify radical expressions including addition, multiplication, and division
4. Rationalize the denominator of a fraction containing a radical in the denominator.

SOLUTIONS to Exercise Set 1.7

2. $(-125)^{-1/3} = \left(-\frac{1}{125}\right)^{1/3} = -1/5.$

5. $\frac{2x^{1/3}}{x^{-3/4}} = 2x^{1/3 + 3/4} = 2x^{13/12}.$

8. $\frac{125^{4/3}}{125^{2/3}} = \frac{5^4}{5^2} = 5^2 = 25.$

11. $\left(\frac{x^{15}}{y^{10}}\right)^{3/5} = \frac{x^9}{y^6}.$

14. $x^{2/3} = \sqrt[3]{x^2}$

17. $(12x^3y^{-2})^{2/3} = \sqrt[3]{\frac{144x^6}{y^4}} = \frac{2x^2}{y^2}\sqrt[3]{18y^2}$

20. $\sqrt[5]{3^2} = 3^{2/5}$

23. $\frac{1}{\sqrt[4]{\frac{4}{9}a^3}} = \frac{1}{(\frac{4}{9}a^3)^{1/4}} = \frac{1}{(\frac{2}{3})^{1/2}a^{3/4}}$

26. $\sqrt{\frac{25}{4}} = \frac{5}{2}$.

29. $\sqrt{(-5)^2} = 5$.

32. $\sqrt{(-\frac{7}{2})^2} = \frac{7}{2}$.

35. Let $x = 2,\ y = 3$.

then $\sqrt{2^2 + 3^2} = \sqrt{13} \neq 2 + 3$.

$\sqrt{13} \neq 5$.

38. 10

41. $\sqrt[3]{y^7} = y^2\sqrt[3]{y}$.

44. $\sqrt{x^5y^4} = x^2y^2\sqrt{x}$.

47. $\sqrt[4]{16x^8y^5} = 2x^2y\sqrt[4]{y}$.

50. $\frac{4}{3\sqrt{11}} \cdot \frac{\sqrt{11}}{\sqrt{11}} = \frac{4\sqrt{11}}{33}$.

53. $\frac{4x^2}{\sqrt{2x}} \cdot \frac{\sqrt{2x}}{\sqrt{2x}} = \frac{4x^2\sqrt{2x}}{2x} = 2x\sqrt{2x}$.

56. $\sqrt[4]{48x^8y^6z^2} = 2x^2y\sqrt[4]{3y^2z^2}$ Note that $48 = 16 \cdot 3$; $\sqrt[4]{16} = 2$.

59. $3\sqrt{x} + 4\sqrt{x} = (3 + 4)\sqrt{x} = 7\sqrt{x}$.

62. $\sqrt{20} - 4\sqrt{45} + \sqrt{80}$

$2\sqrt{5} - 4 \cdot 3\sqrt{5} + 4\sqrt{5}$

$-6\sqrt{5}$

65. $2\sqrt{5} - (3\sqrt{5} + 4\sqrt{5}) = 2\sqrt{5} - 7\sqrt{5} = -5\sqrt{5}$.

68. $\sqrt{8}(\sqrt{2} - \sqrt{3}) = \sqrt{16} - \sqrt{24} = 4 - 2\sqrt{6}$.

71. $(\sqrt{2} - \sqrt{3})^2 = (\sqrt{2} - \sqrt{3})(\sqrt{2} - \sqrt{3}) = 2 - 2\sqrt{6} + 3 = 5 - 2\sqrt{6}$.

74. $(\sqrt[3]{2x}+3)(\sqrt[3]{2x}-3) = (\sqrt[3]{2x})^2 - 9$

$= \sqrt[3]{4x^2} - 9.$

77. $\dfrac{-2}{\sqrt{3}-4}\cdot\dfrac{(\sqrt{3}+4)}{(\sqrt{3}+4)} = \dfrac{-2\sqrt{3}-8}{3-16} = \dfrac{2\sqrt{3}+8}{13}$

80. $\dfrac{4}{2-\sqrt{2y}}\cdot\dfrac{(2+\sqrt{2y})}{(2+\sqrt{2y})} = \dfrac{8+4\sqrt{2y}}{4-2y} = \dfrac{4+2\sqrt{2y}}{2-y}$

83. $\dfrac{(\sqrt{2}+1)(\sqrt{2}+1)}{(\sqrt{2}-1)(\sqrt{2}+1)} = \dfrac{2+2\sqrt{2}+1}{2-1} = 3+2\sqrt{2}.$

86. $\dfrac{2\sqrt{a}}{\sqrt{2x}+\sqrt{y}}\cdot\dfrac{(\sqrt{2x}-\sqrt{y})}{(\sqrt{2x}-\sqrt{y})} = \dfrac{2\sqrt{2ax}-2\sqrt{ay}}{2x-y}$

89. $\sqrt{(-1)(-1)} = \sqrt{(-1)^2} = 1$ by agreement;

Section 1.8 Complex Numbers

BASIC CONCEPTS you should learn from this unit.

1. Definition of $i = \sqrt{-1}$; $i^2 = -1$.
2. Square roots of negative numbers.
3. Expansion of the number system to include numbers of the form $a + bi$. Note that if $b = 0$, we simply have the reals.
4. Adding and subtracting are done by adding (or subtracting) the real parts together and the imaginary parts together:

$$(a + bi) + (x + yi) = a + x + (b + y)i$$

5. Multiplication is done as with real binomials except that when i^2 occurs we replace it with -1.

WHAT YOU SHOULD KNOW HOW TO DO upon completion of this unit.

1. Define complex numbers.
2. Simplify i^n expressions when n is any natural number.
3. Add, subtract, and multiply complex numbers.

SOLUTIONS to Exercise Set 1.8

2. $i^{27} = i^4 \cdot i^4 \cdot i^4 \cdot i^4 \cdot i^4 \cdot i^4 \cdot i^3 = i^3$ Recall that $i^4 = 1$.

$i^3 = i^2 \cdot i = -1$ Since $i^2 = -1$.

5. $-i^{33} = -i$ $i^{32} = \left(i^4\right)^8 = 1^8 = 1$

8. $-i^{39} = -i^3 = -i^2 \cdot i = i$

11. $-\frac{1}{2}$ as a complex number $= -\frac{1}{2} + 0i$

14. $-\sqrt{-5} = -\sqrt{-1} \cdot \sqrt{5} = -i\sqrt{5}$

17. $3 - \sqrt{-49} = 3 - i\sqrt{49} = 3 - 7i$

20. $-0.5 + \sqrt{-32} = -0.5 + \sqrt{-1} \cdot \sqrt{16} \cdot \sqrt{2} = -0.5 + 4i\sqrt{2}$

23. $(3x - 1) + (y + 5)i = 1 - 3i$ means that $3x - 1 = 1$ or $x = 2/3$, and $y + 5 = -3$ or $y = -8$.

26. $(y - 2) + (5x - 3)i = 5 = 5 + 0i$ means that $y - 2 = 5$ or $y = 7$ and $5x - 3 = 0$ or $x = 3/5$

29. Adding the real parts and then the imaginary parts gives $5 + i$.

32. $1\frac{1}{2} - 1\frac{2}{3}i$ or $\frac{3}{2} - \frac{5}{3}i$

35. $-2i(3 + 1) = (-2i)(3) + (-2i)(i)$

$= -6i - 2i^2 = -6i + 2 = 2 - 6i$

38. $(5 + i)(2 - 3i) = 10 - 15i + 2i - 3i^2$

$= 10 - 13i + 3 = 13 - 13i$

41. $(3 - 2i)(2 - i) = 6 - 3i - 4i + 2i^2$

$= 6 - 7i - 2 = 8 - 7i$

44. $x^2 - 2x + 5$ if $x = (2 - i)$, we have

$(2 - i)^2 - 2(2 - i) + 5 = 4 - 4i + i^2 - 4 + 2i + 5 = 5 - 2i + i^2 = 4 - 2i$

47. Prove that $(a + bi) + (c + di) = (c + di) + (a + bi)$.

<u>The left side</u> $= (a + c) + (b + d)i$ and since a,c,b,d are reals, this equals $(c + a) + (d + b)i$ by the commutative property of addition for Reals.

The right side also equals $(c + a) + (d + b)i$ again, by the definition of the addition of complex numbers. But this is exactly the same as the left and the statement is proved.

50. The sum of a number and its additive inverse is zero. So if $(-a - bi) + (a + bi)$ is zero, $-a - bi$ is the additive inverse of $a + bi$.

$$\begin{aligned} -a - bi + (a + bi) &= (-a + a) + (-b + b)i \\ &= 0 + 0i = 0 \end{aligned}$$

and the statement is proved.

SOLUTIONS to Review Exercises.

2. $\{-2\}$

5. F (Natural numbers are integers greater than zero.)

8. $(3a) + (-3a) = 0$ Additive inverse property

11. $9x \cdot 1 = 9x$ the multiplicative identity is 1 .

14.

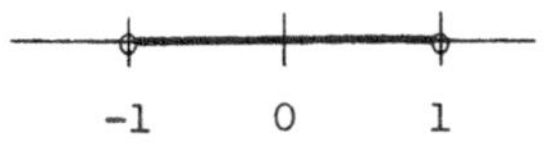

17. Substituting the given values of x and y produces
$(3.25)(12) + (0.15)(80) = 39 + 12 = \$51$

20. -7 is the leading coefficient; the degree is 5.

23. $3x(2x + 1)^2 = 3x(4x^2 + 4x + 1) = 12x^3 + 12x^2 + 3x$

26. $$\begin{aligned} 2a^2 + 3ab + ba + 9b &= a(2a + 3b) + 3(2a + 3b) \\ &= (a + 3)(2a + 3b) \end{aligned}$$

29. $6x^4 + 7x^2 - 3 = (3x^2 - 1)(2x^2 + 3)$

32. $$\frac{\cancel{(a + b)}}{\cancel{(a + 2b)}} \cdot \frac{\cancel{(a + 2b)}(a - 2b)}{\cancel{(a + b)}(a - b)} = \frac{a - 2b}{a - b}$$

35. $$\begin{aligned} x &= x \\ x^2 - x &= x(x - 1) \\ (x - 1)^2 &= (x - 1)(x - 1) \\ \text{LCD} &= x(x - 1)(x - 1) \end{aligned}$$

38. $2 + \frac{4}{a^2 - 4} = \frac{2a^2 - 8}{a^2 - 4} + \frac{4}{a^2 - 4} = \frac{2a^2 - 4}{a^2 - 4}$

41. $x^2 + \frac{\frac{1}{x} + 1}{x - \frac{1}{x}} \cdot \frac{(x)}{(x)} = x^2 + \frac{1 + x}{x^2 - 1} = x^2 + \frac{\cancel{(1 + x)}}{\cancel{(x + 1)}(x - 1)} = x^2 + \frac{1}{x - 1} = \frac{x^3 - x^2 + 1}{x - 1}$

44. $\left(\frac{x^3}{y - 6}\right)^{-4/3} = \frac{x^{-4}}{y^8} = \frac{1}{x^4 y^8}$

47. $\frac{2}{\sqrt{12}} = \frac{2}{\sqrt{4}\sqrt{3}} = \frac{2}{2\sqrt{3}} = \frac{1}{\sqrt{3}} = \frac{1}{3^{1/2}}$

50. $\frac{\sqrt{x}}{(\sqrt{x} + \sqrt{y})} \cdot \frac{(\sqrt{x} - \sqrt{y})}{(\sqrt{x} - \sqrt{y})} = \frac{x - \sqrt{xy}}{x - y}$

53. $(x - 2) + (2y - 1)i = -4 + 7i$ Then $x - 2 = -4$

$x = -2$

and $2y - 1 = 7$

$y = 4$

56. $(2 + i)^2 = 4 + 4i + i^2$

$= 3 - 4i$ since $i^2 = 1$

Chapter 1 - A SHORT QUIZ

1. A set is a ___?___ of objects.
2. Use set notation to describe the set of integers greater than -4 and less than or equal to 2.
3. Is the number $7i$ a rational number?
4. Which property of the real numbers is illustrated by $3xa + 7ya = (3x + 7)a$?
5. If $a < b$, which would be on the left if they were each plotted on a number line?
6. Draw a number line and graph the numbers represented by $-1.5 \leq x \leq 3$.
7. The absolute value of a number is sometimes defined as the ___?___ from zero.
8. State in words the meaning of $-80 < N \leq 1$. (Assume that N is real.)
9. If $a < b$ and $d < 0$, then ad ___?___ bd.

10. Why is $2x^{2/3} + 4x - 1$ NOT a polynomial?

11. What is the degree of $7x^4 - 3x + 5x^5 - 2$?

12. Find the error.

$$(x + 5) - (2x - 2) = x + 5 - 2x - 2 = -x + 3$$

Correct the solution.

13. Evaluate $3x^2 - 2x + 4$ when $x = -2$.

14. Factoring is the process of ____?____ .

15. To factor this expression: $7a^2 - 63b^2$, what would you do first?

16. Factor: $3a^2 - 16a + 21$.

17. A polynomial is said to be ____?____ if it cannot be written as the product of two polynomials, each of positive degree.

18. Find the lowest common denominator for these three fractions.

$$\frac{x}{x^3 - x}, \quad \frac{5}{x}, \quad \frac{x + 3}{(x + 1)^2}$$

19. Divide as indicated.

$$\frac{3x}{x + 2} \div \frac{6x^2}{x^2 - x - 6}$$

20. Simplify this complex fraction.

$$\frac{x - \frac{1}{x}}{2 + \frac{1}{x}}$$

21. Write with a positive exponent.

$$\frac{1}{a^{-2}}$$

22. $x^0 =$ ____?____.

23. Write $\sqrt[3]{x^2}$ in exponent form.

24. $\left(b^{1/n}\right)^n = ?$

25. What is meant by "the principal square root"?

26. Simplify $\frac{1}{\sqrt{x}}$.

27. True or false: $\sqrt{9} + \sqrt{36} = \sqrt{45}$?

28. True or false: $\sqrt{9} \cdot \sqrt{36} = \sqrt{324}$?

29. Find the sum of $(3 - 2i) + (-6 - 7i)$.

30. Find the product of $i(3 - 2i)$.

Quiz Answers Chapter 1

1. collection

2. $\{-3,-2,-1,0,1,2\}$

3. No, it is complex.

4. The distributive property

5. a would be to the left of b.

6. 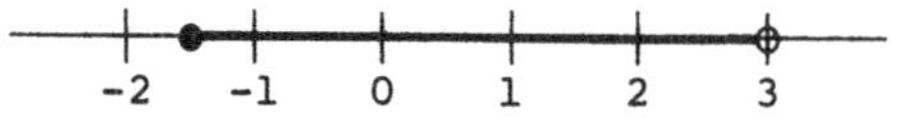

7. distance

8. N is greater than -80 and less than or equal to 1.

9. $>$

10. A rational exponent of $\frac{2}{3}$ is not allowed.

11. 5

12. The right side should be $x + 5 - 2x + 2 = -x + 7$.
(arrow under "+ 2": error)

13. 20

14. Finding the numbers or expressions that, when multiplied, yield the given number or expression.

15. Factor out the common factor, 7.

16. $(3a - 7)(a - 3)$

17. prime or irreducible

18. $x(x - 1)(x + 1)^2$

19. $\frac{x - 3}{x}$

20. $\frac{x^2 - 1}{2x + 1}$

21. a^2

22. 1

23. $x^{2/3}$

24. b

25. The positive square root is the principal square root.

26. $\frac{\sqrt{x}}{x}$

27. False

28. True

29. -3 - 9i

30. 2 + 3i

CHAPTER 2

Equations and Inequalities

Section 2.1 Linear Equations in One Unknown

BASIC CONCEPTS you should learn from this unit.

1. A linear equation in one unknown is a statement of equality containing one variable that has the exponent one (1); for example, $3x \cdot 2 = 7$.

2. A solution to an equation is a numeric value of the variable that makes the statement true.

3. Linear equations can be solved by adding a number or expression or subtracting the same number or expression from both sides of the equation, and/or by multiplying or dividing both sides of the equation by a number or expression.

4. A linear equation is often called a first degree equation.

5. If it is necessary to multiply or divide a linear equation by an expression containing the unknown, the resulting equation may not be equivalent to the original one; hence, the solution must be checked in the original equation.

WHAT YOU SHOULD KNOW HOW TO DO upon completion of this unit.

1. Recognize linear equations.

2. Solve linear equations.

3. Verify supposed solutions.

4. Determine whether an equation is an identity or a conditional equation.

SOLUTIONS to Exercise set 2.1

2. $3 \cdot \frac{5}{2} - 4 = \frac{15}{2} - \frac{8}{2} = \frac{7}{2} = \frac{5}{2}$ so $\frac{5}{2}$ is not a solution.

5. $3x + 5 = -1, \quad 3x = -6, \quad x = -2$

8. $\frac{1}{2}s + 2 = 4, \quad \frac{1}{2}s = 2, \quad s = 4$

11. $0 = -\frac{1}{2}a - \frac{2}{3}, \quad \frac{1}{2}a = -\frac{2}{3}, \quad a = -\frac{4}{3}$

14. $2x - 1 = 3x + 2, \quad 2x - 3x = 2 + 1, \quad x = -3$

17. $2(3b + 1) = 3b - 4, \quad 6b + 2 = 3b - 4, \quad 3b = -6, \quad b = -2$

20. $-3(x - 2) = 2(x + 4), \quad -3x + 6 = 2x + 8, \quad -5x = 2, \quad x = -\frac{2}{5}$

23. $-4(2x + 1) - (x - 2) = -11, \quad -8x - 4 - x + 2 = -11, \quad -9x - 2 = -11, \quad -9x = -9, \quad x = 1$

26. $8 - 2kx = -3x, \quad 8 = 2kx - 3x, \quad 8 = x(2k - 3), \quad x = \frac{8}{2k - 3}$

29. $\frac{x}{2} = \frac{5}{3}$ Multiply both sides by 6

$3x = 10$

$x = \frac{10}{3}$

32. $\frac{5}{a} - \frac{3}{2} = \frac{1}{4}$ (LCD = a), $4a\left(\frac{5}{a}\right) - 4a\left(\frac{3}{2}\right) = 4a\left(\frac{1}{4}\right)$, $20 - 6a = a$, $20 - 7a$, $a = \frac{20}{7}$

35. $\frac{1}{x-2} + \frac{1}{2} = \frac{2}{x-2}$ multiplying by the LCD, which is $2(x - 2)$, gives

$2(x-2)\left[\frac{1}{x-2} + \frac{1}{2} = \frac{2}{x-2}\right] = 2 + x - 2 = 4$, $x = 4$,

check:

$\frac{1}{4-2} + \frac{1}{2} \stackrel{?}{=} \frac{2}{4-2}$, $\frac{1}{2} + \frac{1}{2} \stackrel{?}{=} \frac{2}{2}$, $1 = 1$

so that the solution is correct.

38. $\frac{3}{x-1} + \frac{2}{x+1} = \frac{5}{(x+1)(x-1)}$ LCD = $(x + 1)(x - 1)$

$(x+1)(x-1)\left[\frac{3}{x-1} + \frac{2}{x+1} = \frac{5}{(x+1)(x-1)}\right]$; $3x + 3 + 2x - 2 = 5$; $5x + 1 = 5$;

$5x = 4$; $x = \frac{4}{5}$

41. $\frac{4}{b} - \frac{1}{b+3} = \frac{3b+2}{(b+3)(b-1)}$ LCD = $b(b + 3)(b - 1)$

Multiplying both sides by the LCD gives

$4b^2 + 8b - 12 - b^2 + b = 3b^2 + 2b$

$7b = 12$

$b = \frac{12}{7}$

This, with a bit of work, does check!

44. $\frac{2x-1}{x-5} + 3 = \frac{3x-2}{5-x}$ <u>or</u> $\frac{2x-1}{x-5} + 3 = \frac{2-3x}{x-5}$

Therefore, LCD = $x - 5$; multiplying, we have

$2x - 1 + 3x - 15 = 2 - 3x$; $8x - 16 = 2$; $8x = 18$; $x = \frac{9}{4}$. This also checks.

47. $2x + 1 = 3x - 1$; $2 = x$. Thus it is a conditional equation.

50. $5(x - 1) = 10$

$x - 1 = 2$ Divide by 5

$x = 3$ Add 1.

Yes, all are equivalent.

53. $3(x^2 + 2x + 1) = -6$

$x^2 + 2x + 1 = -2$ Divide by 3

$(x + 1)^2 = -2$ Simply factoring the left-hand side

Yes, all are equivalent.

Section 2.2 Applications

BASIC CONCEPTS you should learn from this unit.

1. "Applied" mathematics problems generally involve word sentences as well as numbers, so we have to "translate" such problems into mathematics. Helpful steps in this procedure are the following.
   - A. Read the problem carefully.
   - B. Assign a letter to represent the quantity to be found.
   - C. Write an equation that reflects the relationships stated in the problem.
   - D. Apply the rules for solving equations.

   (This sequence sounds easy, but many students make it hard by not doing the first two steps carefully.)

WHAT YOU SHOULD KNOW HOW TO DO upon completion of this unit.

1. Translate word statements into an algebraic equation.
2. Solve the equation using the methods of Section 2.1.
3. Substitute your answer in the original problem to check whether it makes sense.
4. Solve literal equations (often formulas) for a specified letter.

SOLUTIONS to Exercise Set 2.2

2. The number of sedans = n, so the number of station wagons would be $3n - 20$.

5. Let n = Fred's age now

   Then $n + 12$ = John's age now

   Also $n - 4$ = Fred's age 4 years ago

   $n + 12 - 4 = n + 8$ = John's age 4 years ago

   Equation: $n + 8 = 2(n - 4)$

8. n = larger number

   $n - 5$ = smaller number

   $n + (n - 5) = 11$

Section 2.2 Applications

11. n = short piece

$n + 4$ = longer piece

The sum should be 12 meters.

$n + (n + 4) = 12$

14. Choose n = number of \$10 bills

Then $2n + 4$ = number of \$5 bills

and $n - 6$ = number of \$20 bills

The equation found by the <u>value</u> of the bills should equal \$460.

$$10n + 5(2n + 4) + 20(n - 6) = 460$$

17. Let n = number of \$6 tickets

$2n$ = number of \$5 tickets

$3n + 1$ = number of \$3 tickets

The equation found by the value of the tickets should equal \$503.

$$6n + 5(2n) + 3(3n + 1) = 503$$

20. Let n = amount invested in black and white film

$4000 - n$ = amount invested in color film

$0.12n$ = profit from black and white film

$0.21(4000 - n) = 0.12n - 150$

23. The distance traveled by each truck is, of course, equal when the faster truck overtakes the slower one. Using $d = rt$ as the formula for each truck and then setting these expressions equal, we have

t = time of the faster truck

$t - 2$ = time of the slower truck

$55t = 50(t - 2)$

distance = distance

Also, t will be the time required for the second truck to overtake the first.

27. Draw a diagram:

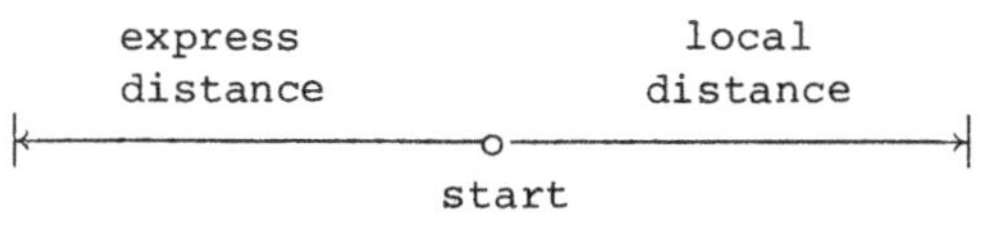

Use $d = rt$.

Let r = speed of the local

$2r$ = speed of the express

Then

$$\underset{\text{local distance}}{4r} + \underset{\text{express distance}}{4(2r)} = \underset{\text{total distance}}{480}$$

29. Let n = number of ounces of Ceylon tea

$8 - n$ = number of ounces of Formosa tea

Equation found by value of tea:

$$1.50n + 2.00(8 - n) = 1.85(8)$$

Ceylon value + Formosa value = mixture value

32. Let n = the number of kilograms of green beans. (This, again, is a "value" equation.)

$$1 \cdot n + 1.30(100) + 1.40(90) = 1.20(190 + n)$$

bean value + corn value + pea value = mix value

35. Let t = the number of hours with both computers.

$$\frac{t}{4} + \frac{t}{6} = 1$$

part by A + part by B = total job

38. Let t = the number of hours for hose B alone.

$$\frac{2}{3} + \frac{3}{t} = 1$$

part of fill done by A + part of fill done by B = total fill

41. Let r = speed of the plane in still air

$r + 20$ = speed with the wind

$r - 20$ = speed against the wind

Times are equal in this problem, so use $t = \frac{d}{h}$.

$$\frac{300}{r - 20} = \frac{400}{r + 20}$$

44. $V = \frac{1}{3}\pi r^2 h;\quad 3V = \pi r^2 h;\quad \frac{3V}{\pi r^2} = h$

47. $A = \frac{1}{2}h(b + b');\quad 2A = h(b + b');\quad \frac{2A}{h} = b + b';$ hence, $b = \frac{2A}{h} - b'$ or $b = \frac{2A - b'h}{h}$.

Section 2.3 Linear Inequalities

BASIC CONCEPTS you should learn from this unit.

1. A Linear Inequality is an inequality containing a variable to the first power.

2. The solution of a linear inequality is the set of all values of the variable that make the statement true.

3. Graphs of linear inequalities are generally intervals on a number line.

4. Procedures for solving an inequality are the same as those for solving equations, with this one exception: multiplying or dividing by a negative number reverses the sense of the inequality.

WHAT YOU SHOULD KNOW HOW TO DO upon completion of this unit.

1. Recognize a linear inequality.

2. GRAPH an inequality on a number line.

3. Solve an inequality algebraically, using addition, subtraction, multiplication. and division.

SOLUTIONS to Exercise Set 2.3

2. $-4 < x \leq 1$: $(-4, 1]$

5. $-12 \leq x \leq -3$: $[-12, -3]$

8. $x < 17$: $(-\infty, 17)$

11. $[5, 8]$: $5 \leq x \leq 8$

14. $[-3, 10]$: $-3 \leq x \leq 10$

17. $[0, \infty)$: $x \geq 0$

20. $x + 5 < 4$

$x + 5 - 5 < 4 - 5$

$x < -1$

23. $x - 3 \geq 2$

$x \geq 2 + 3$

$x \geq 5$

26. $-5 > b - 3$

$-5 + 3 > b$

$-2 > b$

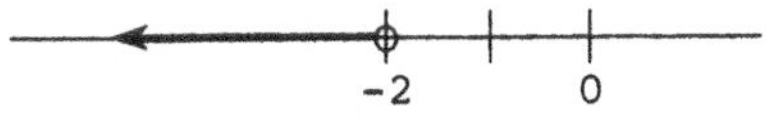

29. $2x \geq 0$

$x \geq 0$

32. $3x - 2 > 4$

$3x > 6$

$x > 2$

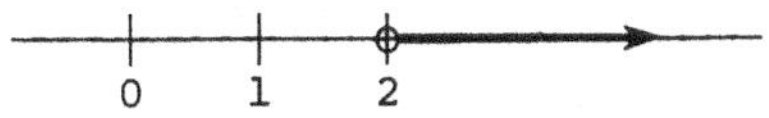

35. $\frac{4}{5 - 3x} < 0$

For this inequality to be true, 5-3x must be negative or

$5 - 3x < 0$

$5 < 3x$

$\frac{5}{3} < x$

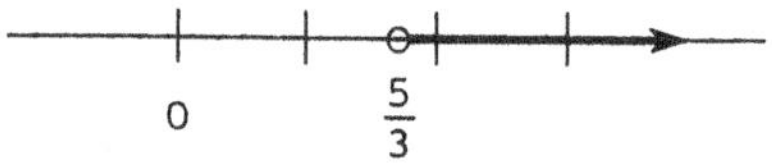

38. $\frac{1}{2}y - 2 \leq 2$

$\frac{1}{2}y \leq 4$

$y \leq 8$: $(-\infty, 8]$

41. $4(2x + 1) < 16$

$8x + 4 < 16$

$8x < 12$

$x < 3/2$: $\left(-\infty, \frac{3}{2}\right)$

44. $4(x - 3) \geq 3(x - 2)$

$4x - 12 \geq 3x - 6$

$x \geq 6$: $[6, \infty)$

47. $\frac{2}{3}(x + 1) + \frac{5}{6} \geq \frac{1}{2}(2x - 1) + 4$

$\frac{2}{3}x + \frac{2}{3} + \frac{5}{6} \geq x - \frac{1}{2} + 4$ Multiply both sides by the LCD, 6.

$4x + 4 + 5 \geq 6x - 3 + 24$

$4x + 9 \geq 6x + 21$

$-12 \geq 2x$

$x \leq -6$: $(-\infty, -6]$

50. $\frac{x}{5} - \frac{1 - x}{2} > \frac{x}{2} - 3$ Multiply both sides by 10.

$2x - 5 + 5x > 5x - 30$

$2x > -25$

$x > -25/2$: $\left(\frac{-25}{2}, \infty\right)$

53. $-2 < 4x \leq 5$

$-\frac{1}{2} < x \leq \frac{5}{4}$: $\left(-\frac{1}{2}, \frac{5}{4}\right]$

56. $5 \le 3x - 1 \le 11$

$6 \le 3x \le 12$ Add 2

$2 \le x \le 4$ Divide by 3

[2, 4]

59. $-8 < 2 - 5x \le 7$

$-10 \le -5x \le 5$ subtract 2

$2 \ge x \ge -1$ Divide by -5

or

$-1 \le x \le 2$. [-1, 2]

62. A is to be less than B:

$160 < 100 + 0.20m$

$60 < 0.20m$

$300 < m$ or more than 300 miles

65. Let n = number of copies sold

$12.50n > 19000 + 6n$

income from n copies > cost of n copies

$6.5n > 19000$

$n > 2923$ 2924 copies or more

Section 2.4 Absolute Value in Equations and Inequalities

BASIC CONCEPTS you should learn from this section.

1. Definition of Absolute Value: $|x| = \begin{cases} x & \text{if } x \ge 0 \\ -x & \text{if } x < 0 \end{cases}$

2. Solving equations and inequalities with Absolute Value expressions is done by using your understanding of absolute value <u>and</u> the rules learned from the preceding units for solving equations and inequalities. For example,

$|3x + 1| = 5$

$3x + 1 = 5$, $x = \frac{4}{3}$

$-(3x + 1) = 5$, $3x + 1 = -5$, $x = -2$

$|3x + 1| > 5$

$3x + 1 > 5$, $x > \frac{4}{3}$

$-(3x + 1) > 5$, $3x + 1 > -5$, $x > -2$

3. Graphs of equations or inequalities with Absolute Value generally consist of one or two intervals on a number line.

WHAT YOU SHOULD KNOW HOW TO DO upon completion of this unit.

1. Define Absolute Value.

2. Solve equations and inequalities containing Absolute Value expressions.

3. Graph equations and inequalities that include Absolute Value

SOLUTIONS to Exercise Set 2.4

2. $|r - 5| = \frac{1}{2}$; $r - 5 = \frac{1}{2}$ $\quad -(r - 5) = \frac{1}{2}$

$\underline{r = 5\frac{1}{2}}$, $\quad \underline{r = 4\frac{1}{2}}$

5. $|-3x + 1| = 5$

$-3x + 1 = 5$ or $+3x - 1 = 5$

$-3x = 4$ $\quad 3x = 6$

$\underline{x = -\frac{4}{3}}$ $\quad \underline{x = 2}$

8. $\frac{1}{|x|} = 5$ Therefore, $1 = 5|x|$ or $|x| = \frac{1}{5}$.

So $\underline{x = \frac{1}{5}}$ or $\underline{x = -\frac{1}{5}}$

11. $|x + 1| > 3$

$x + 1 > 3$ or $-(x + 1) > 3$

$x > 2$ $\quad x + 1 < -3$

$x < -4$

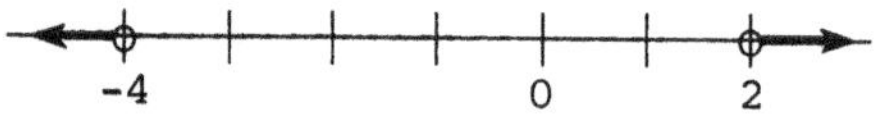

12. $|3x + 6| \leq 12$

$3x + 6 \leq 12$ $\quad -3x - 6 \leq 12$

$3x \leq 6$ $\quad -3x \leq 18$

$x \leq 2$ $\quad x \geq -6$

14. Look out! Absolute Value cannot be negative! Therefore only values that satisfy $|3x + 2| \geq 0$ will do. What we're really saying here is that $|3x + 2| = -1$ is an invalid statement.

17. $|x - 3| \geq 4$

$x - 3 \geq 4$ $\quad -(x - 3) \geq 4$

$x \geq 7$ $\quad x - 3 \leq -4$

$x \leq -1$

$(-\infty, -1]$ or $[7, \infty)$

20. $\frac{|3x + 2|}{2} \le 4$

$|3x + 2| \le 8$

$3x + 2 \le 8$

$3x \le 6$

$x \le 2$

$-(3x + 2) \le 8$

$3x + 2 \ge -8$

$3x \ge -10$

$x \ge -\frac{10}{3}$

$[-\frac{10}{3}, 2]$

23. $\left|\frac{5 - x}{3}\right| > 4$:

$\frac{5 - x}{3} > 4$

$5 - x > 12$

$x < -7$

$(-\infty, -7)$

or

$\frac{x - 5}{3} > 4$

$x - 5 > 12$

$x > 17$

$(17, \infty)$

26. Prove that $|x|^2 = x^2$.

For $x < 0$

$|x| = -x$

$(-x)^2 = x^2$

For $x = 0$

$|x| = x$

$x^2 = x^2$

For $x > 0$

$|x| = x$

$x^2 = x^2$

28. Let x = weekly income.

$|x - 300| \le 50$

(difference between weekly income and \$300) is less than or equal to (\$50)

Section 2.5 The Quadratic Equation

BASIC CONCEPTS you should learn from this unit.

1. Definition of a QUADRATIC EQUATION as one that can be written in the form $ax^2 + bx + c = 0$ where a, b, c are real numbers, but $a \neq 0$.

2. The methods of solving Quadratic Equations are

   A. Simply taking a square root;

   B. Factoring, then using the Zero Product Principle, that is, if $a \cdot b = 0$, either $a = 0$ or $b = 0$;

   C. Completing the square;

   D. Using the Quadratic Formula, $x = \frac{-b \pm \sqrt{b^2 - 4ac}}{2a}$

3. Quadratic equations have <u>two</u> roots which can be real or complex.

4. The <u>discriminant</u>, $(b^2 - 4ac)$, tells us the nature of the roots:

$b^2 - 4ac > 0$	$b^2 - 4ac = 0$	$b^2 - 4ac < 0$
two real roots	double root	two complex roots

5. Certain equations with radicals and exponents, although <u>not quadratic</u>, can be solved by the foregoing methods if a proper substitution of variable is made or by using the squaring process.

WHAT YOU SHOULD KNOW HOW TO DO upon completion of this unit.

1. Recognize an equation as quadratic.

2. Solve quadratic equations by various methods.

3. Solve certain types of equations that are quadratic in form.

SOLUTIONS to Exercise Set 2.5

2. $4x^2 - 64 = 0$, $4x^2 = 64$, $x = \pm 4$

5. $5y^2 - 25 = 0$, $5y^2 = 25$, $y = \pm\sqrt{5}$

8. $(4t + 1)^2 - 3 = 0$, $(4t + 1)^2 = 3$, $4t + 1 = \pm\sqrt{3}$, $t = \frac{-1\pm\sqrt{3}}{4}$

11. $x^2 - 3x + 2 = 0$, $(x - 2)(x - 1) = 0$; $x - 2 = 0$ and $x - 1 = 0$, so $x = 2, 1$.

14. $3r^2 - 4r + 1 = 0$, $(3r - 1)(r - 1) = 0$, $3r - 1 = 0$ and $r - 1 = 0$, so $r = \frac{1}{3}, 1$.

17. $y^2 - 4y = 0$, $y(y - 4) = 0$, $y = 0$, $y = 4$

20. $2s^2 - 5s - 3 = 0$, $(2s + 1)(s - 3) = 0$, $s = -1/2$, $s = 3$

23. $6x^2 - 5x + 1 = 0$, $(3x - 1)(2x - 1) = 0$; $x = 1/3$, $x = 1/2$

26. $t^2 - 2t = 15$

$$t^2 - 2t + 1 = 15 + 1$$
$$(t - 1)^2 = 16$$
$$t - 1 = \pm 4$$
$$t = 5$$
$$t = -3$$

29. $3x^2 + 8x = 3$

$$x^2 + \frac{8}{3}x + \frac{16}{9} = 1 + \frac{16}{9}$$
$$\left(x + \frac{4}{3}\right)^2 = \frac{25}{9}$$
$$x + \frac{4}{3} = \pm\frac{5}{3}$$
$$x = \frac{1}{3}, -3$$

32. $3x^2 - 4x = -3$

$$x^2 - \frac{4}{3}x + \frac{4}{9} = -1 + \frac{4}{9}$$

$$\left(x - \frac{2}{3}\right)^2 = -\frac{5}{9}$$

$$x - \frac{2}{3} = \frac{\pm i\sqrt{5}}{3} \; ; \quad x = \frac{2 \pm i\sqrt{5}}{3}$$

35. $3x^2 + 2x = -1$

$$x^2 + \frac{2}{3}x + \frac{1}{9} = -\frac{2}{9}$$

$$\left(x + \frac{1}{3}\right)^2 = -\frac{2}{9}$$

$$x + \frac{1}{3} = \pm\frac{i\sqrt{2}}{3} \; ; \quad x = \frac{-1 \pm i\sqrt{2}}{3}$$

38. $2x^2 + 3x + 3 = 0$

$$x = \frac{-3 \pm \sqrt{9 - 24}}{4}$$

$$= \frac{-3 \pm i\sqrt{15}}{4}$$

41. $5y^2 - 4y + 5 = 0$

$$y = \frac{4 \pm \sqrt{16 - 100}}{10} = \frac{4 \pm \sqrt{-84}}{10}$$

$$= \frac{2 \pm i\sqrt{21}}{5}$$

44. $2x^2 + 4x - 3 = 0$

$$x = \frac{-4 \pm \sqrt{16 + 24}}{4} = \frac{-4 \pm \sqrt{40}}{4}$$

$$= \frac{-2 \pm \sqrt{10}}{2}$$

47. $4u^2 + 3u = 0$

$$u = \frac{-3 \pm \sqrt{9 - 0}}{8}$$

$$= \frac{-3 \pm 3}{8} \qquad \underline{u = 0}\,, \quad \underline{u = -\frac{3}{4}}$$

50. $2t^2 + 2t + 3 = 0$

$$t = \frac{-2 \pm \sqrt{4 - 24}}{4} = \frac{-1 \pm i\sqrt{5}}{2}$$

53. $2x^2 + 5x + 4 = 0$

$$x = \frac{-5 \pm \sqrt{25 - 32}}{4} = \frac{-5 \pm i\sqrt{7}}{4}$$

56. $x^2 + 2 = 0 \quad x^2 = -2 \quad x = \pm\sqrt{-2} = \pm i\sqrt{2}$

59. $a^2 + b^2 = c^2 \quad b^2 = c^2 - b^2 \quad b = \pm\sqrt{c^2 - b^2}$

62. $A = \pi r^2 \quad r^2 = \frac{A}{\pi} \quad r = \pm\sqrt{\frac{A}{\pi}}$ or $\frac{\pm\sqrt{A\pi}}{\pi}$

65. $x^2 - 2x + 3 = 0$ Therefore, $b^2 - 4ac = 4 - 12 = -8$.
two complex roots

68. $2x^2 + x + 5 = 0$; $b^2 - 4ac = 1 - 10 = -9$
two complex roots

71. $3x^2 + 2x = 0$, $b^2 - 4ac = 4$
two real rational roots

74. $3x^2 + x - 5 = 0$, rearrange it, then $b^2 - 4ac = 1 + 60 = 61$
there are two real irrational roots

77. $3r^2 - 6r + 1 = 0$, $b^2 - 4ac = 6 - 12 = -6$
there are two complex roots

80. $4s^2 + 4s + 1 = 0$, $b^2 - 4ac = 16 - 16 = 0$
there is a double root (real).

83. $b^2 - 4ac$ is to be less than 0, so
$2x^2 + 3x + k = 0 \quad b^2 - 4ac = 3 - 8k$
and $3 - 8k < 0$, $\underline{k > 3/8}$

86. $x - \sqrt{13 - x} = 1$
$x - 1 = \sqrt{13 - x}$
$x^2 - 2x + 1 = 13 - x$
$x^2 - x - 12 = 0 \quad (x - 4)(x + 3) = 0 \quad x = 4$
Solution $x = -3$ will not check.

89. $\sqrt{3x + 4} - \sqrt{2x + 1} = 1$

$$\sqrt{3x + 4} = 1 + \sqrt{2x + 1}$$
$$3x + 4 = 1 + 2\sqrt{2x + 1} + 2x + 1$$
$$x + 2 = 2\sqrt{2x + 1}$$
$$x^2 + 4x + 4 = 4(2x + 1)$$
$$x^2 - 4x = 0$$
$$x = 0, \quad x = 4$$

Both roots check.

92. $\sqrt{5x + 1} + \sqrt{4x - 3} = 7$

$$\sqrt{5x + 1} = 7 - \sqrt{4x - 3}$$
$$5x + 1 = 49 - 14\sqrt{4x - 3} + 4x - 3$$
$$x - 45 = -14\sqrt{4x - 3}$$
$$x^2 - 90x + 2025 = 196(4x - 3)$$
$$= 784x - 588$$
$$x^2 - 874x + 2613 = 0$$
$$(x - 3)(x - 871) = 0$$

Isn't that a dandy?

$x = 3$ checks.

$x = 871$ doesn't check.

95. $\frac{6}{x^2} + \frac{1}{x} - 2 = 0$; $6 + x - 2x^2 = 0$; $(3 + 2x)(2 - x) = 0$

$x = -\frac{3}{2}$ $x = 2$

Substitute $u = \frac{1}{x}$.

$$6u^2 + u - 2 = 0$$
$(3u + 2)(2u - 1) = 0$, $u = -\frac{2}{3}$ or $x = -\frac{3}{2}$

$u = \frac{1}{2}$ or $x = 2$

98. $3x^{4/3} - 4x^{2/3} - 4 = 0$

Let $u = x^{2/3}$; then

$$3u^2 - 4u - 4 = 0$$
$$(3n + 2)(u - 2) = 0$$

$u = -\frac{2}{3}$ $u = +2$

$x^{2/3} = -\frac{2}{3}$ $x^{2/3} = 2$

$\underline{x = \left(-\frac{8}{27}\right)^{1/2}}$ $x = 8^{1/2} = \underline{\sqrt{8}}$

101. Prove that if r_1 and r_2 are roots to a quadratic equation, then

$$r_1 r_2 = \frac{c}{a} \quad \text{and} \quad r_1 + r_2 = -\frac{b}{a}$$

Proof:

If r_1 and r_2 are roots to a quadratic equation, then

$$r_1 = \frac{-b + \sqrt{b^2 - 4ac}}{2a} \quad \text{and} \quad r_2 = \frac{-b - \sqrt{b^2 - 4ac}}{2a}$$

$$r_1 r_2 = \left(\frac{-b + \sqrt{b^2 - 4ac}}{2a}\right)\left(\frac{-b - \sqrt{b^2 - 4ac}}{2a}\right)$$

$$= \frac{b^2 - b^2 + 4ac}{4a^2} = \frac{c}{a} \quad \underline{\text{as required}}$$

also

$$r_1 + r_2 = \frac{-b + \sqrt{b^2 - 4ac}}{2a} + \frac{-b - \sqrt{b^2 - 4ac}}{2a}$$

$$= \frac{-2b}{+2a} = -\frac{b}{a} \quad \underline{\text{as required}}$$

Section 2.6 Applications - Quadratic Equations

BASIC CONCEPTS you should learn from this unit.

1. Procedure for solving these word problems is similar to that for other word problems: Read carefully, choose a letter to represent the unknown, form an equation, and solve it.

2. Be alert to meaningless solutions in the context of the problem (negative age, for example).

3. Use of drawings is often helpful.

WHAT YOU SHOULD KNOW HOW TO DO upon completion of this unit.

1. Set up an equation of second degree to solve an applied problem.

2. Solve word problems similar to those illustrated and check your answers.

SOLUTIONS to Exercise Set 2.6

2. Let x = number of days for the designer

then $x + 16$ = number of days for the assistant

$$\frac{6}{x} + \frac{6}{x + 16} = 1 \qquad \text{Multiply both sides by } x(x + 16) \text{ gives}$$

$$6x + 96 + 6x = x^2 + 16x$$

$$x^2 + 4x - 96 = 0$$

$$(x + 12)(x - 8) = 0$$

$$x = -12, \qquad \text{Reject this root.}$$

$$x = 8 \text{ days}$$

$$x + 16 = 24 \text{ days}$$

5.

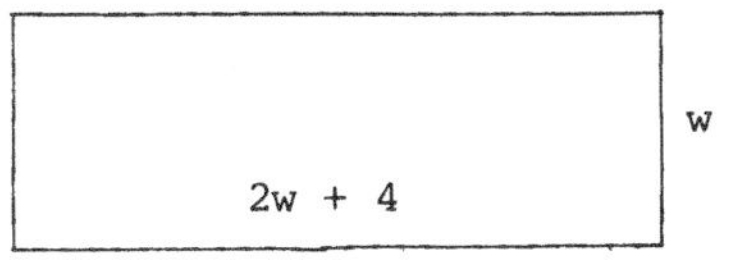

w = width

$2w + 4$ = length

area = $w(2w + 4)$

$2w^2 + 4w = 48$

$w^2 + 2w - 24 = 0$

$(w + 6)(w - 4) = 0$ $\cdot$ $w = 4'$

($w = -6$; reject this root) $2w + 4 = 12'$

8.

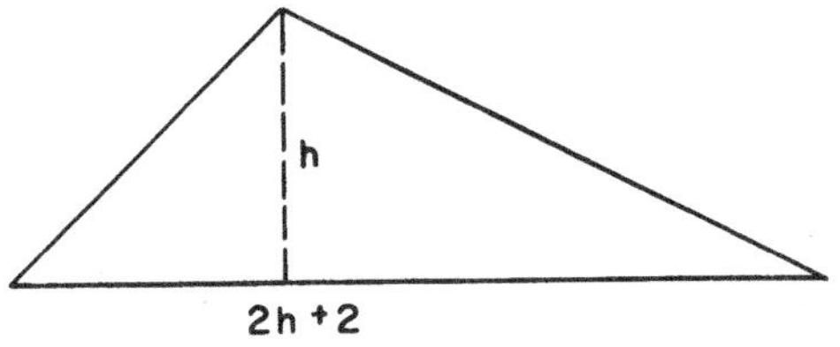

h = height

$2h + 2$ = base

$A = \frac{1}{2}Bh$

$\frac{1}{2}(2h + 2) \cdot h = 12$

$(h + 1)h = 12$

$h^2 + h - 12 = 0$

$(h + 4)(h - 3) = 0$ $h = 3'$

base = 8'

11. Let x = the number

$x + \frac{1}{x} = \frac{26}{5}$; $5x^2 + 5 = 26x$, $5x^2 - 26x + 5 = 0$, $(5x - 1)(x - 5) = 0$

so $\underline{x = \frac{1}{5}}$ or $\underline{x = 5}$

14. n = first odd number

$n + 2$ = next consecutive odd number

then $\frac{1}{n} + \frac{1}{n + 2} = \frac{8}{15}$ LCD = $15n(n + 2)$

$15n + 30 + 15n = 8n^2 + 16n$

$8n^2 - 14n - 30 = 0$

$4n^2 - 7n - 15 = 0$

$(4n + 5)(n - 3) = 0$

hence $n = 3$ or $n = -\frac{5}{4}$ Reject this root

and $n + 1 = 5$

17. Let n = number of shares

cost of each share = $\frac{1200}{n}$

$\frac{1200}{n - 30} = \frac{1200}{n} + 2$ LCD = $n(n - 30)$

$1200n = 1200n - 36000 + 2n^2 - 60n$

$2n^2 - 60n - 36000 = 0$

$n^2 - 30n - 18000 = 0$

$(n - 150)(n + 120) = 0$ Therefore, $\underline{n = 150}$ shares

20. x = daily rate of pay

$\frac{480}{x}$ = number of days worked at \$x per day

$\frac{480}{x} = \frac{480}{x - 8} + 2$ LCD = x(x - 8)

$480x - 3840 = 480x + 2x^2 - 16x$

$2x^2 - 16x + 3840 = 0$; $x^2 - 8x + 1920 = 0$

$(x - 48)(x + 40) = 0$

hence, x = \$48 per day

Section 2.7 Second Degree Inequalities

BASIC CONCEPTS you should learn from this unit.

1. A SECOND DEGREE inequality has the form $ax^2 + bx + c \leq 0$ or $ax^2 + bx + c \geq 0$ or can be put in that form.

2. If factorable, second degree inequalities can be solved by analyzing the signs of the factors.

3. Inequalities of the form $\frac{ax + b}{cx + d} < 0$ can also be solved by analysis of signs.

WHAT YOU SHOULD KNOW HOW TO DO upon completion of this unit.

1. Recognize second degree inequalities.

2. Solve second degree inequalities by factoring and sign analysis.

3. Solve inequalities of the form $\frac{ax + b}{cx + d} < 0$ by sign analysis.

SOLUTIONS to Exercise Set 2.7

2. $x^2 + 3x - 4 \leq 0$

$(x + 4)(x - 1) \leq 0$

```
x + 4 ---0+++++++++++++++++++++++++++++++++++++
x - 1 ----------------------0+++++++++++++++
```

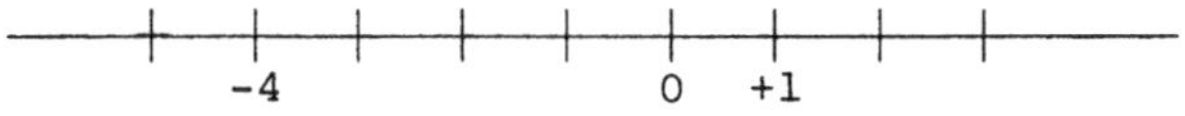

Signs are unlike when $-4 < x < 1$

final solution $-4 \leq x \leq 1$

5. $4x - 2x^2 < 0$

$2x(2 - x) < 0$

```
    2x -----------0+++++++++++++++++++++++++++
 2 - x +++++++++++++++++++++0----------------
```

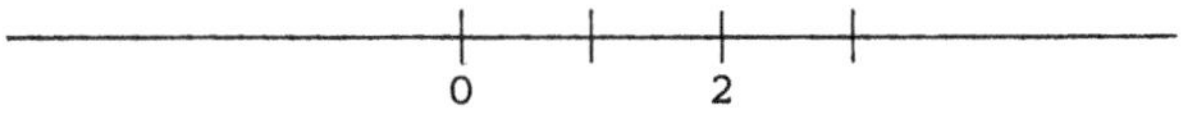

Signs are unlike when $x < 0$ or $x > 2$

8. $\dfrac{x - 6}{x + 4} \geq 0$

(x - 6) ------------------------------0++++++

(x + 4) ----0++++++++++++++++++++++++++++++++++

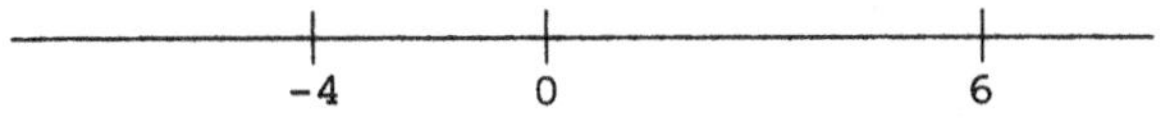

Solution when signs are the same $x < -4$ or $x \geq 6$

<u>or</u> $x - 6 = 0$

11. $\dfrac{3s + 2}{2s - 1} \geq 0$

3s + 2 ---------------0+++++++++++++++++++++

2s - 1 ----------------------0++++++++++++++

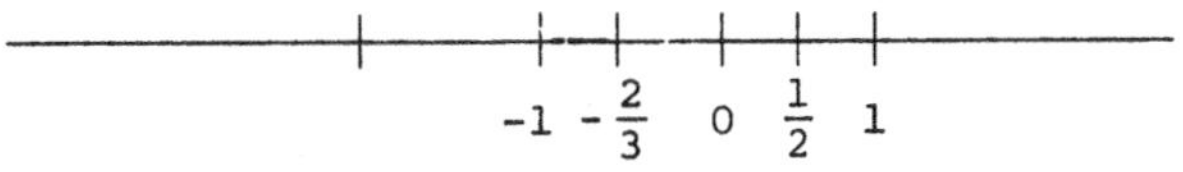

Solution: $s > \frac{1}{2}$ or $s \leq -\frac{2}{3}$

14. $(x - 4)(2x + 5)(2 - x) \leq 0$

x - 4 ---------------------------------0++

2x + 5 -------0+++++++++++++++++++++++++++

2 - x ++++++++++++++++++++++++0----------

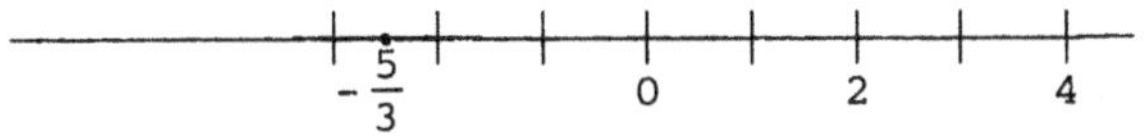

For $(x - 4)(2x + 5)(2 - x) \leq 0$ <u>one</u> or all three of the factors must be negative, <u>or</u> any one of the factors must be zero. Thus

$x \geq 4$ or $-5/2 \leq x \leq 2$

17. $2x^2 - 3x - 5 < 0$

$(2x - 5)(x + 1) < 0$

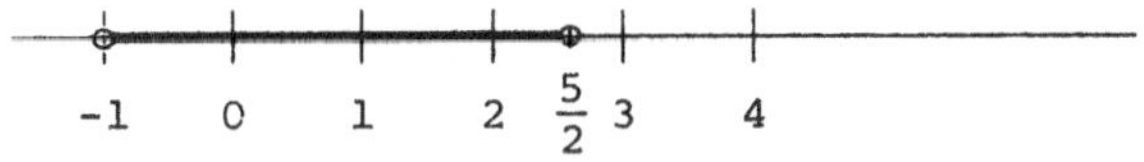

20. $\dfrac{3x + 2}{2x - 3} \geq 0$

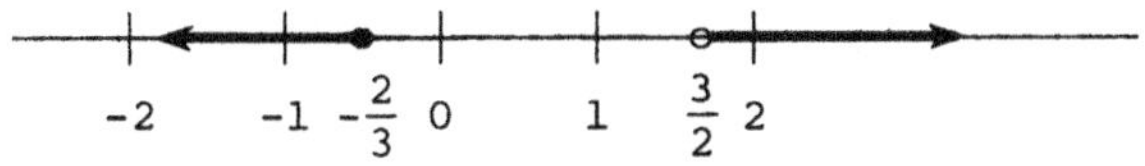

23. $6x^2 + 8x + 2 \geq 0$

$2(3x^2 + 4x + 1) \geq 0$

$2(3x + 1)(x + 1 \geq 0$

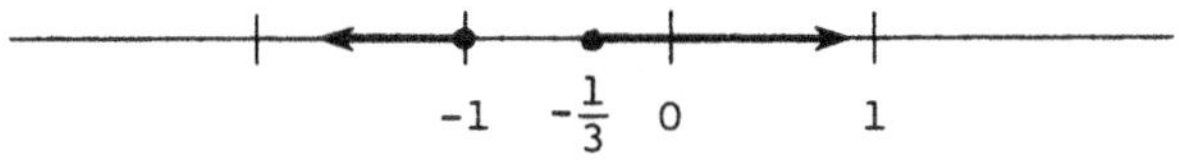

26. $(2x + 5)(3x - 2)(x + 1) < 0$

```
(2x + 5)  ---0++++++++++++++++++++++++++++++
(3x - 2)  ----------------------0+++++++++++
 (x + 1)  -----------0++++++++++++++++++++++
```

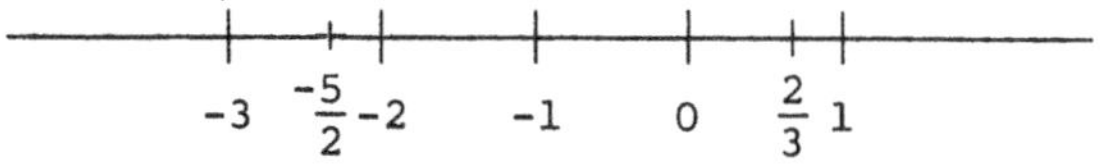

Solution

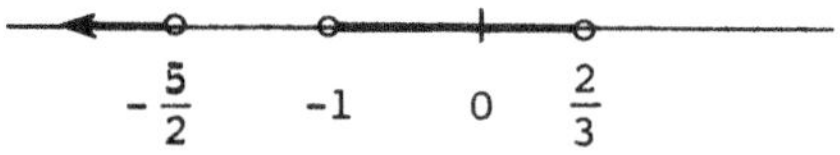

29. $\sqrt{(x - 2)(x + 1)}$ will be real when $(x - 2)(x + 1) \geq 0$;

that is, when $x \leq -1$ or $x \geq 2$.

32. $\sqrt{2x^2 + 3x + 1} = \sqrt{(2x + 1)(x + 1)}$ is real for

$x \leq -1 \quad x \geq -\frac{1}{2}$

SOLUTIONS to Review Exercises Chapter 2

2. $2(2x - 3) - 3(x + 1) = -9$

$4x - 6 - 3x - 3 = -9$

$x = 0$

5.

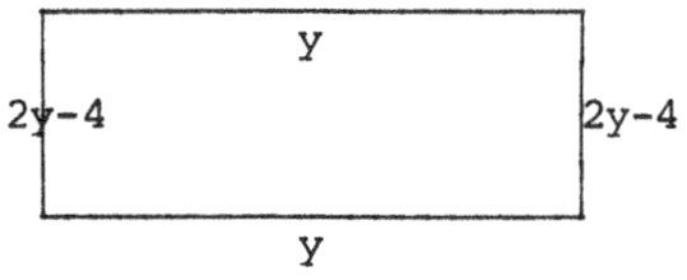

Let y = length

$y + y + 2y - 4 + 2y - 4 = 12$

$6y - 8 = 12$

$6y = 20 \qquad y = 10/3$

$2y - 4 = 8/3$

8. Let b = number of hours for B to do job alone

$\frac{2}{3} + \frac{2}{b} = 1$

$2b + 6 = 3b$

$6 = b$ Answer: 6 hours

11. $3 \leq 2x + 1$

$2 \leq 2x$

$x \geq 1$

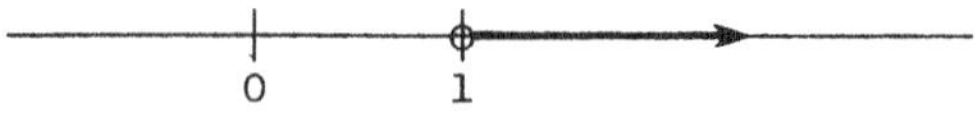

14. $\frac{-1}{2x - 5} < 0$ True only if $2x - 5 > 0$ or when $x > 5/2$

$\left(\frac{5}{2}, \infty\right)$

17. $|4x - 1| = 5$ $\qquad -(4x - 1) = 5$

$4x - 1 = 5$ $\qquad 4x - 1 = -5$

$x = 6/4 = 3/2$ $\qquad 4x = -4$

$\underline{x = 3/2}$ $\qquad \underline{x = -1}$

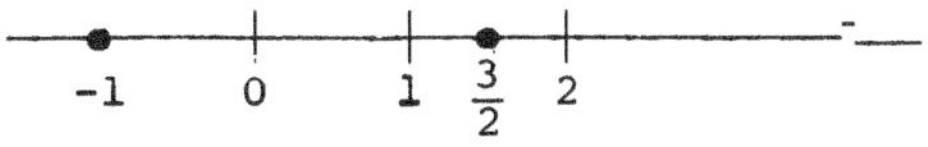

20. $|3x - 2| \geq 6$ $\qquad -(3x - 2) \geq 6$

$3x - 2 \geq 6$ $\qquad 3x - 2 \leq -6$

$3x \geq 8$ $\qquad 3x \leq -4$

$x \geq 8/3$ $\qquad x \leq -4/3$

Solution: $\left(-\infty, -\frac{4}{3}\right]$, $\left[\frac{8}{3}, \infty\right)$

23. $x^2 - 2x + 6 = 0$

$x^2 - 2x + 1 = -6 + 1$

$(x - 1)^2 = -5$

$x - 1 = \pm i\sqrt{5}$; $\qquad x = 1 \pm i\sqrt{5}$

26. $49x^2 - 9 = 0$

$49x^2 = 9$

$7x = \pm 3$ $\qquad x = \frac{3}{7}, -\frac{3}{7}$

29. $3r^2 = 2r + 5$

$3r^2 - 2r - 5 = 0$ $\qquad b^2 - 4ac = 4 + 60 = 64$

Therefore, there are two real rational roots.

32. $x + \sqrt{2x - 3} = 4$

$\sqrt{2x - 3} = 4 - x$

$2x - 3 = 16 - 8x + x^2$

$x^2 - 10x + 19 = 0$ $\qquad$ Therefore, $x = \frac{10 \pm \sqrt{24}}{2} = \underline{5 \pm \sqrt{6}}$

35. $\left(1 - \frac{2}{x}\right)^2 - 8\left(1 - \frac{2}{x}\right) + 15 = 0$

let $u = 1 - 2/x$ so that $u^2 - 8u + 15 = 0$

$(u - 5)(u - 3) = 0$

$u = 5$ $\qquad u = 3$

$5 = 1 - \frac{2}{x}$ $\qquad 3 = 1 - \frac{2}{x}$

$x = -\frac{1}{2}$ $\qquad x = -1$

38. $x^2 + 4x - 5 \geq 0$

$(x + 5)(x - 1) \geq 0$ when $x \leq -5$ or $x \geq 1$

Interval notation: $(-\infty, -5]$ or $[1, \infty)$

Chapter 2 - A SHORT QUIZ

1. The exponent on the variable in a linear equation is ___?___

2. A solution to a linear equation is ___?___ ___?___

3. Which of the following are linear equations?

   (a) $2x - 7 = \frac{1}{y}$ (b) $4 - x = 7x$ (c) $5x^2 + x = 1$

4. Is $\frac{7}{3}$ a solution of $3x - 5 = 2$?

5. The sum of two consecutive odd integers is 76. Write an equation that will allow you to find the integers.

6. Solve $\frac{1}{R} = \frac{1}{a} + \frac{1}{b}$ for b.

7. If $a > b$ and $m < 0$, place the correct inequality symbol between am ___?___ bm .

8. Graph the inequality $-3 \leq x < 4$ on a number line.

9. Write this interval as an inequality: $[-3, 5)$.

10. $\frac{-5}{3x - 7} < 0$ only when $3x - 7$ is ___?___.

11. For what values of x is $|3x - 2| = -1$?

12. Write an absolute value inequality expressing this graph:

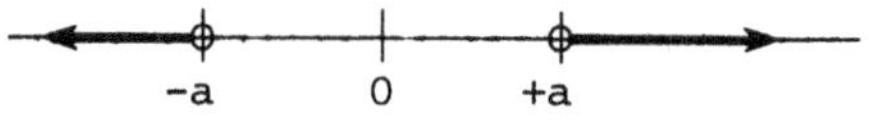

13. $ax^2 + bx + c = 0$ is called the general quadratic equation; the constants a, b, and c are real with what restrictions?

14. In the quadratic formula $b^2 - 4ac$ is called the ___?___ and when it is negative we know that the roots are ___?___ .

15. Under what special condition does a quadratic equation have two identical roots?

16. Write a quadratic equation having 1 and -7 as roots.

17. Solve by completing the square:

   $3x^2 + 2x + 1 = 0$

18. Two integers differ by 5 and their product is 150. What are they?

19. A man can do a job in a hours and his son can do it in b hours. How long will it take them together?

20. Given $x^2 + 4x - 21 < 0$, what would be your first step in finding the solution?

21. Given $(x - 4)(x + 2)(x + 5) > 0$. Using a sign analysis diagram, determine the solution.

22. Under what condition of $3x + 5$ will the following be true: $\frac{3x + 5}{x - 7} = 0$?

Quiz Answers Chapter 2

1. 1 (one)

2. a number value of the variable that makes the statement (equation) true.

3. b

4. Yes

5. $n + n + 2 = 76$

6. $b = \frac{Ra}{a - b}$

7. $<$

8. 

9. $-3 \leq x < 5$

10. positive

11. $\emptyset$

12. $|x| > a$

13. $a \neq 0$

14. discriminant, complex

15. Discriminant = 0 .

16. $(x - 1)(x + 7) = 0$ <u>or</u> $x^2 + 6x - 7 = 0$

17. $3x^2 + 2x + 1 = 0$

$$x^2 + \frac{2}{3}x + \frac{1}{9} = -1 + \frac{1}{9}$$

$$(x + \frac{1}{3})^2 = -\frac{8}{9}$$

$$x + \frac{1}{3} = \frac{\pm 2i\sqrt{2}}{3}$$

$$x = \frac{-1 \pm 2i\sqrt{2}}{3}$$

18. 10, 15 also -10, -15

19. $\frac{ab}{a + b}$ hours

20. Factor the left-hand side.

21.
```
x - 4 ----------------------------0+++++++
x + 2 -----------0++++++++++++++++++++++++
x + 5 --0+++++++++++++++++++++++++++++++++

-------|-----------|-----|------------|--------
      -5          -2     0            4
```

Answer: $-5 < x < -2$; and $x > 4$

22. only if $3x + 5 = 0$

CHAPTER 3

Functions

Section 3.1 Rectangular Coordinate Systems

BASIC CONCEPTS you should learn from this unit.

1. The rectangular coordinate system consists of two perpendicular number lines, a horizontal one, called the x-axis, and a vertical one, called the y-axis.

2. Each point in the coordinate plane can be located by two coordinates (x, y), each of which is a real number.

3. The Distance Formula $d = \sqrt{(x_2 - x_1)^2 + (y_2 - y_1)^2}$ provides an easy way to obtain the distance between two points in the plane.

4. The graph of an equation in two variables is a line in the plane whose points are found by constructing a table of x and y values from the equation and plotting those pairs of coordinate values.

5. Symmetry prevails

   A. with respect to the y-axis if $-x$ replacing x leaves the equation unchanged;

   B. with respect to the x-axis if $-y$ replacing y leaves the equation unchanged;

   C. with respect to the origin if both of the foregoing replacements leave the equation unchanged.

WHAT YOU SHOULD KNOW HOW TO DO upon completion of this unit.

1. Plot points on a set of coordinate axes.

2. Find the distance between two points in the plane.

3. Determine whether three points are collinear.

4. Sketch graphs of equations in two variables.

5. Determine the symmetry of the graph of an equation without graphing.

SOLUTIONS to Exercise Set 3.1

2.

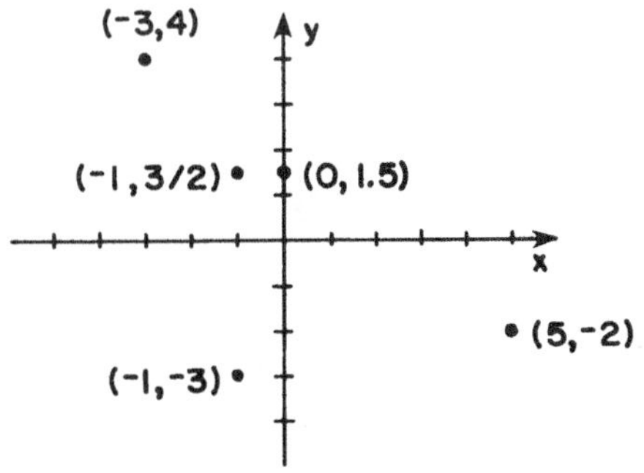

5. $P_1(-1, -5)$, $P_2(-5, -1)$

$$d = \sqrt{(-4)^2 + (4)^2} = \sqrt{32} = 4\sqrt{2}$$

8. $P_1(-\frac{1}{2}, 3)$, $P_2(-1, -\frac{3}{4})$

$$d = \sqrt{\left(-\frac{1}{2}\right)^2 + \left(-\frac{15}{4}\right)^2} = \sqrt{\frac{1}{4} + \frac{225}{16}} = \frac{\sqrt{229}}{4}$$

11. $R\left(-1, \frac{1}{2}\right)$, $S\left(-\frac{3}{2}, 1\right)$, $T(2, -1)$

$$\overline{RS} = \sqrt{\frac{1}{2}^2 + \left(-\frac{1}{2}\right)^2} = \sqrt{\frac{1}{2}} = \frac{\sqrt{2}}{2}$$

$$\overline{RT} = \sqrt{\frac{3}{2}^2 + 3^2} = \sqrt{\frac{9}{4} + 9} = \frac{\sqrt{45}}{2}$$

$$\overline{ST} = \sqrt{2^2 + \left(\frac{7}{2}\right)^2} = \sqrt{4 + \frac{49}{4}} = \frac{\sqrt{65}}{2}$$

Thus $\overline{RS}$ is the shortest side.

14. Let $A(2, -3)$, $B(-1, -1)$, $C(3, 4)$

Then

$$\overline{AB}^2 = 2^2 + 3^2 = 13$$

$$\overline{AC}^2 = 7^2 + 1^2 = 50 \qquad 13 + 41 \neq 50$$

$$\overline{BC}^2 = 5^2 + 4^2 = 41$$

Therefore, ABC is <u>not</u> a right triangle.

17. $A(-1, 2)$, $B(1, 1)$, $C(5, -1)$

$$\overline{AB} = \sqrt{1^2 + 2^2} = \sqrt{5} \qquad \overline{AB} + \overline{BC} = \overline{AC}$$

$$\overline{AC} = \sqrt{3^2 + 6^2} = 3\sqrt{5} \qquad \text{yes, collinear}$$

$$\overline{BC} = \sqrt{2^2 + 4^2} = 2\sqrt{5}$$

20. $A(-1, 5)$, $B(1, 1)$, $C(-2, -8)$

$$\overline{AB} = \sqrt{6^2 + 2^2} = 2\sqrt{10}$$

$$\overline{AC} = \sqrt{3^2 + 1^2} = \sqrt{10} \qquad \underline{\text{not}}\ \text{collinear}$$

$$\overline{BC} = \sqrt{9^2 + 3^2} = \sqrt{100} > 10$$

23. A(9, 2), B(11, 6), C(3, 5), D(1, 1)

$\overline{AB} = \sqrt{4^2 + 2^2} = 2\sqrt{5}$

$\overline{CD} = \sqrt{4^2 + 2^2} = 2\sqrt{5}$ — equal

$\overline{AD} = \sqrt{1^2 + 8^2} = \sqrt{65}$

$\overline{BC} = \sqrt{1^2 + 8^2} = \sqrt{65}$ — equal

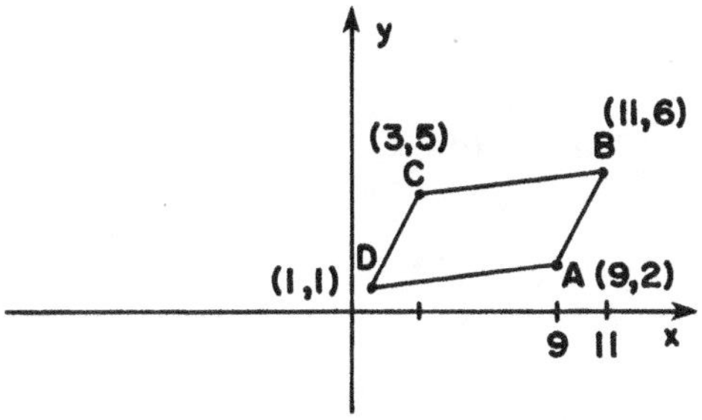

Yes, the points are the vertices of a parallelogram.

26. Plotting points to see locations:

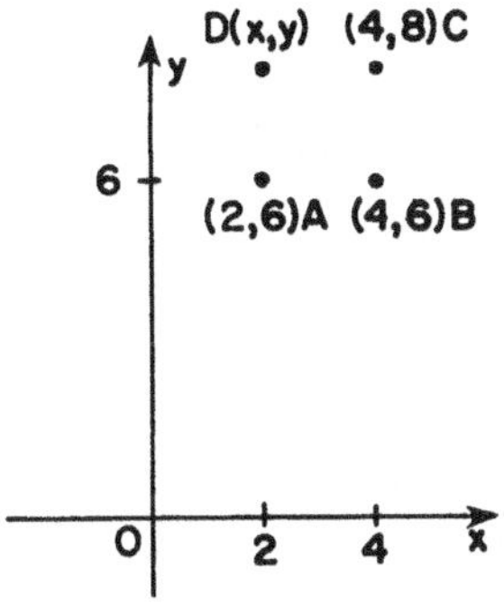

It would seem that (x, y) should be (2, 8). Verify by the distance formula.

29. $y = \sqrt{x}$

x	0	1	4	9	16
y	0	1	2	3	4

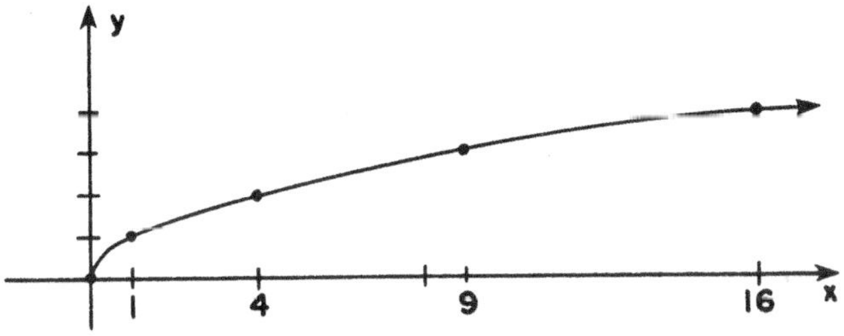

32. $y = 2 - |x|$

x	0	2	-2	3	-3	5	-5
y	2	0	0	-1	-1	-3	-3

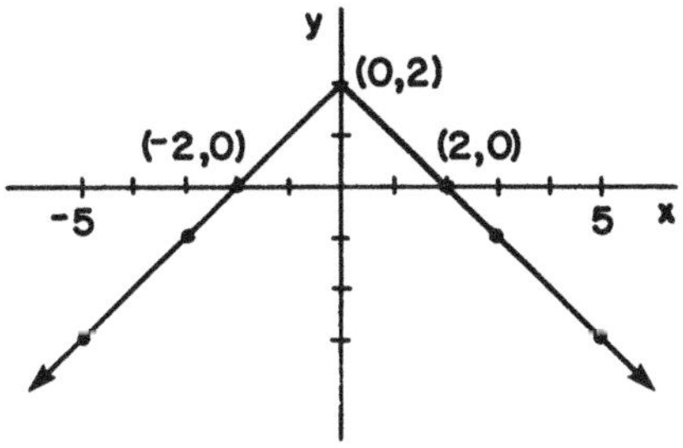

35. $y = x^3 + 1$

Substitution of $(-x)$ for x or $(-y)$ for y reveals no symmetry with either axis.

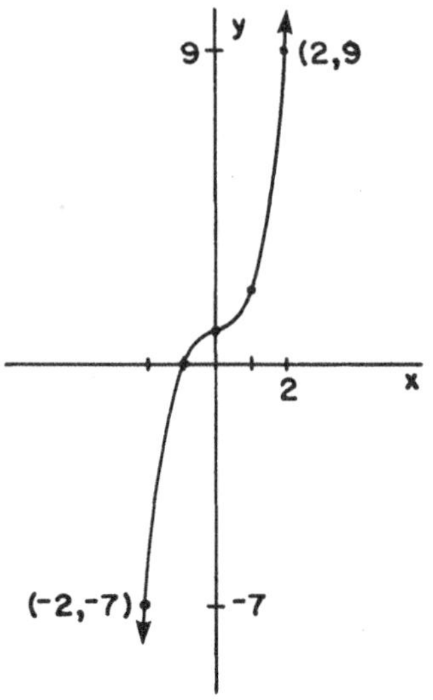

38. $y = 3x$

Substitution of $-x$ and $-y$ leaves the equation unchanged. Therefore, there is symmetry with the origin.

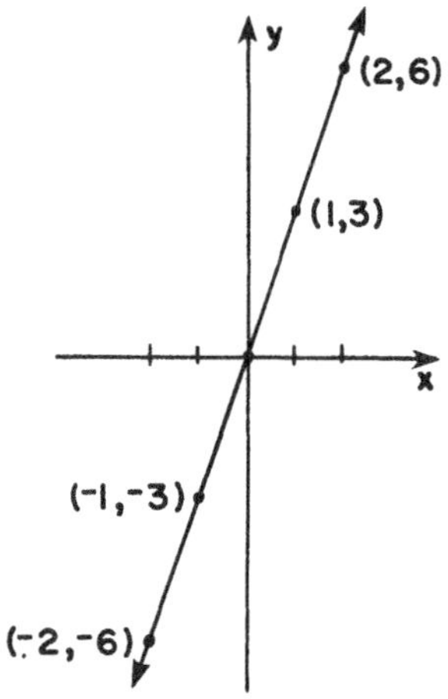

41. $y^2 = x - 4$

$-x$ for x CHANGES equation

$-y$ for y DOES NOT CHANGE equation

Therefore, there is only x-axis symmetry.

44. $y = (x - 2)^2$

Substituting $-x$ for x changes the equation, and substituting $-y$ for y changes the equation.

Both substitutions at once also change the equation.

Therefore, there is no symmetry with the axes or origin.

47. $y^2x + 2x^2 = 4x^2y$

Substituting $-x$ for x changes the first term to negative;

substituting $-y$ for y changes the last term to negative.

Both substitutions at once yield $-y^2x + 2x^2 = -4x^2y$, which is not the same as the original equation.

Therefore, there is no symmetry.

52. $4x^2 + 9y^2 = 36$: Since both x and y are squared and since $(-x)^2 = x^2$ and $(-y)^2 = y^2$, we have symmetry with respect to the x-axis, the y-axis, and the origin.

Section 3.2 Functions and Function Notation

BASIC CONCEPTS you should learn from this unit.

1. Definition of a function as a rule that assigns exactly one y-value to each x-value.
2. Definition of Domain as the set of all x-values.
3. Definition of Range as the set of all y-values.
4. If a vertical line intersects the graph of an equation at only one point, the equation represents a function.
5. Restrictions on the Domain are often required:
   A. x may not cause a denominator to be zero.
   B. x may not cause an even root (square root, fourth root, etc.) to be negative.

   We are dealing here only with Real numbers.
6. Domain restrictions will sometimes but not always cause restrictions in range.
7. The function notation $f(x)$ is read "f of x."

   In $f(x) = 3x + 7$, the $f(x)$ is simply a convenient way of denoting the rule.

   $f(2)$ in this case is $3 \cdot 2 + 7 = 13$ and is called the Function Value when $x = 2$.

WHAT YOU SHOULD KNOW HOW TO DO upon completion of this unit.

1. Define function, Domain, and Range.
2. Determine the Domain and Range of a given function.
3. Determine if a given equation is or is not a function.
4. Evaluate a function, that is, find $f(x)$ if given an x value.

SOLUTIONS to Exercise Set 3.2

2. $y = x^2 + x \qquad -2 \le x \le 1$

x	y
-2	2
-1	0
0	0
1	2
-1/2	-1/4

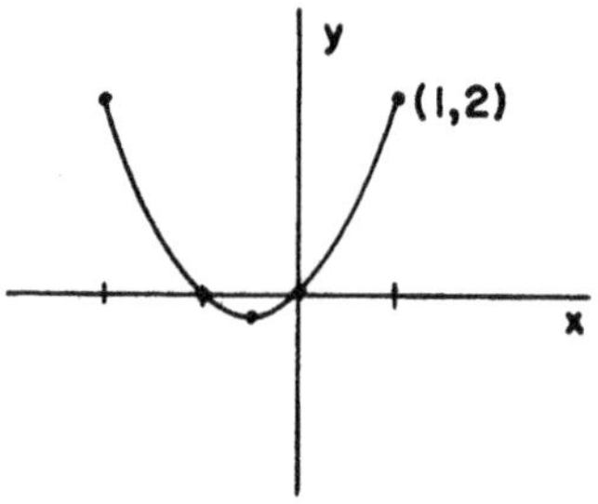

the Range is: $-\frac{1}{4} \le y \le 2$

5. $y = \sqrt{x - 1}$

x	y
1	0
2	1
5	2

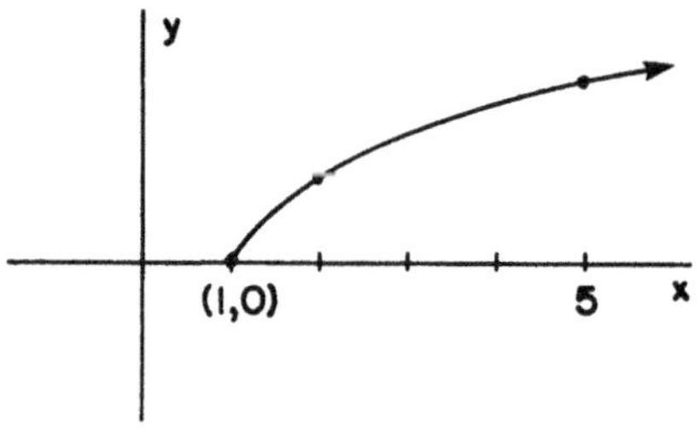

the Range is: $y \ge 0$

8. $f(x) = \sqrt{5 - x}$ the Domain is: $x \le 5$
($x > 5$ causes a negative radicand.)

11. $f(x) = \dfrac{\sqrt{x - 1}}{x - 2}$ the Domain is: $x \ge 1$ so that $x - 1 \ge 0$
but $x \neq 2$ (zero denominator)

14. $f(x = x^2$ in order that $f(x) = 2$ we need to choose $x = \pm\sqrt{2}$.

17. $f(x) = 2x^2 + 5$
$f(0) > 2 \cdot 0 + 5 = \underline{5}$

20. $f(3x) = 2(3x)^2 + 5 = \underline{18x^2 + 5}$

23. $g(-3) = (-3)^2 + 2(-3) = 3$

26. $g(-x) = (-x)^2 + 2(-x) = x^2 - 2x$

29. Use a calculator $\dfrac{(-2.73)^2 + 1}{3(-2.73) - 1} = \dfrac{8.4529}{-7.19} \approx -1.176$

32. $f(-x) = \frac{(-x)^2 + 1}{3(-x) - 1} = \frac{x^2 + 1}{-3x - 1}$

35. $r(-8.27) = \frac{-8.27 - 2}{(-8.27)^2 + 2(-8.27) - 3} = \frac{-10.27}{48.8529} = -0.210$

38. $2r(a) = 2 \cdot \frac{a - 2}{a^2 + 2a - 3} = \frac{2a - 4}{a^2 + 2a - 3}$

41. $I = prt$ Here $P = x$ dollars, $r = 7\%$, $t = 4$

$I(x) = 0.07 \cdot 4 \cdot x$ or $I(x) = 0.28x$.

44. Since the perimeter, P, of a square is 4 times one side and the area is a side squared, if the area is A, then one side is $\sqrt{A}$ and $P = 4\sqrt{A}$.

Section 3.3 Graphs of Functions

BASIC CONCEPTS you should learn from this unit.

1. The graph of a function is the graph of $y = f(x)$. For example, if $f(x) = -2x + 7$ we would graph $y = -2x + 7$.

2. There are ways to describe functions other than by equations, for example, as a table of paired values.

3. A function is increasing if $f(x_2) > f(x_1)$ when $x_2 > x_1$ and decreasing if $f(x_2) < f(x_1)$ when $x_2 > x_1$.

4. A constant function will have the same value for all allowable x.

5. All polynomials are functions.

WHAT YOU SHOULD KNOW HOW TO DO upon completion of this unit.

1. Graph a given function.

2. Determine intervals when a function is increasing, decreasing, or constant.

3. Given a relationship between two variables in words, write a function statement of the variables.

SOLUTIONS to Exercise Set 3.3

2. $f(x) = 3 - 2x$

decreasing on all x

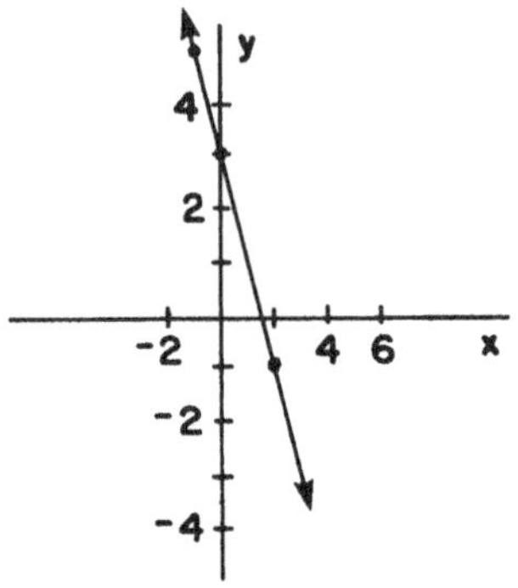

5. $f(x) = 9 - x^2$

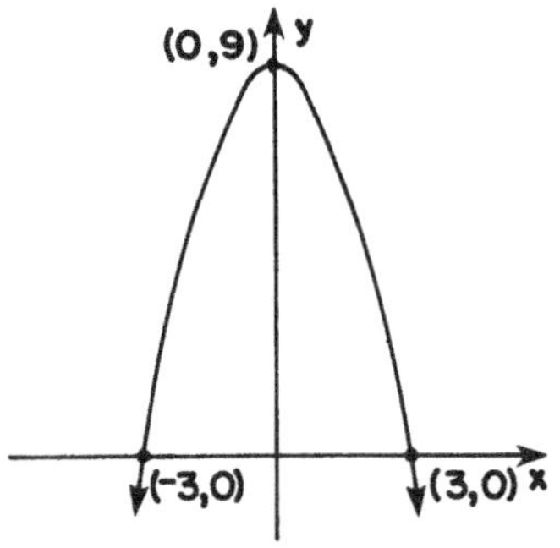

increasing: $x < 0$

decreasing: $x > 0$

8. $f(x) = |1 - x|$

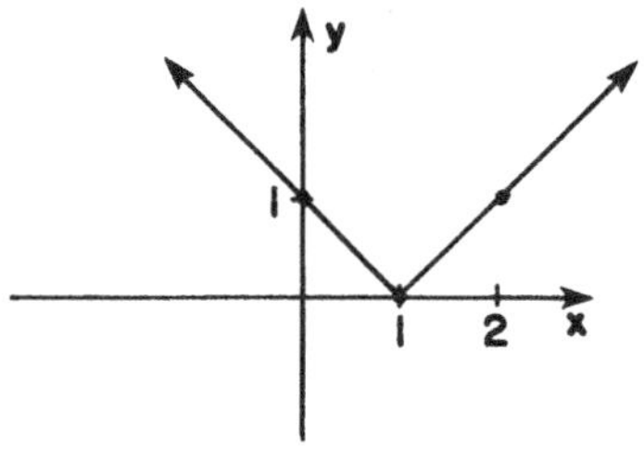

increasing: $x > 1$

decreasing: $x < 1$

11. $f(x) = \begin{cases} x, & x < 2 \\ 2, & x \geq 2 \end{cases}$

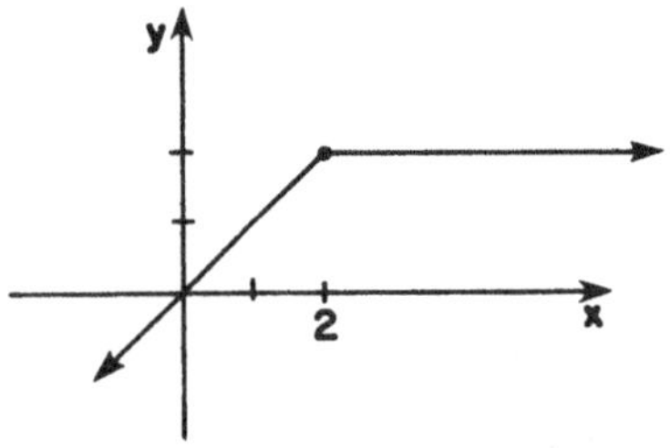

increasing: $x < 2$

never decreasing

constant: $x > 2$

14. $f(x) = \begin{cases} 2 & \text{if } x \text{ is an integer} \\ -1 & \text{if } x \text{ is NOT an integer} \end{cases}$

$f(x)$ always a constant (either 2 or -1)

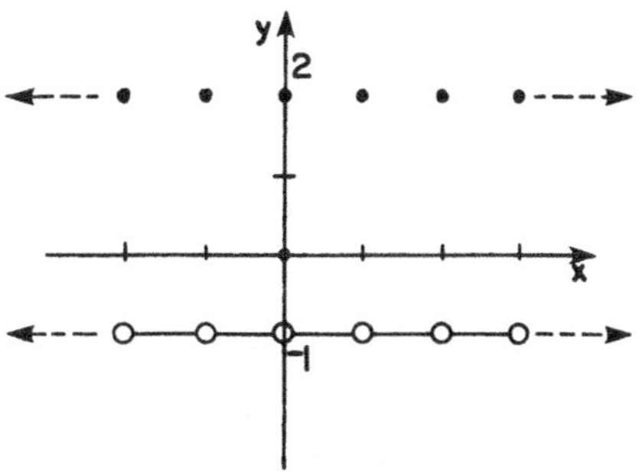

17. $f(x) = x^2$

$g(x) = 2x^2$

$h(x) = \frac{1}{2}x^2$

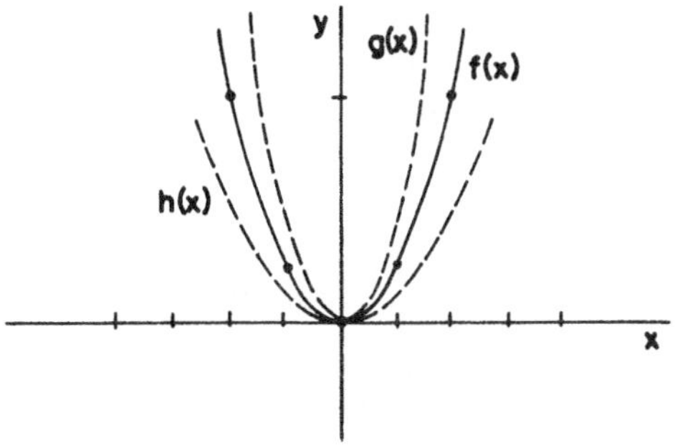

20. $f(x) = x^2 - 2$

$g(x) = 2 - x^2$

x	f(x)	x	g(x)
0	-2	0	2
2	2	2	-2
-2	2	-2	-2

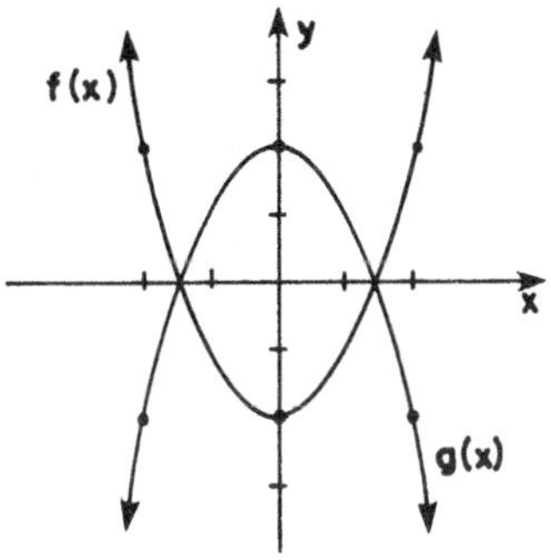

23. $f(x) = x^3$

$g(x) = -x^3$

Use table of values.

x	f(x)	x	g(x)
0	0	0	0
1	1	1	-1
-1	-1	-1	1

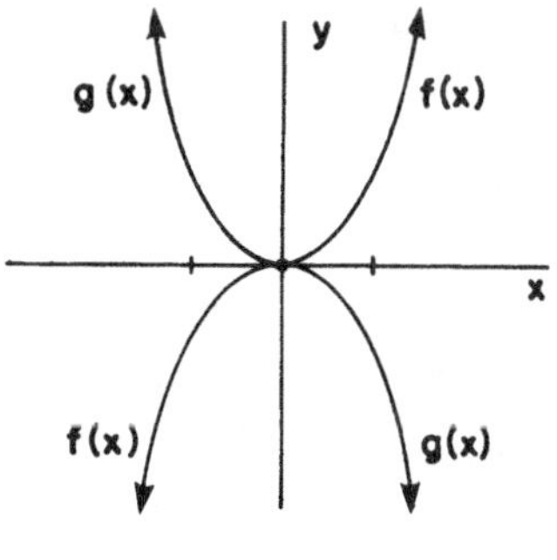

26. The Function will have three parts, corresponding to the three salary groups. Dues = f(s) with s = exmployees salary.

$$f(s) = \begin{cases} 60 & \text{if } s < 8000 \\ 60 + 0.01s & \text{if } 8000 \leq s < 15000 \\ 130 + 0.02s & \text{if } s \geq 15000 \end{cases}$$

29. m = number of miles

(a) C(m) = cost function

C(m) = \$14 + 8¢ per mile

Therefore, $C(m) = 14 + 0.08m$

(b) DOMAIN = allowable values of m, $m > 0$

(c) $C(100) = 14 + 0.08(100) = \$22$

Section 3.4 Linear Functions

BASIC CONCEPTS you should learn from this unit.

1. The slope of a line is defined as the ratio

$$\frac{y_2 - y_1}{x_2 - x_1}$$

for any two points $P_1(x_1, y_1)$ and $P_2(x_2, y_2)$ on the line.

2. The point-slope form of the straight line is $y - y_1 = m(x - x_1)$ where m is the slope and (x_1, y_1) are the coordinates of a point on the line.

3. The point (0, b) on a straight line is called the y-intercept, since it is where the line crosses the y-axis.

4. $y = mx + b$ is the slope-intercept form of the straight line.

5. The general first degree equation is $Ax + By + C = D$ and solving for y yields $-\frac{A}{B}x - \frac{C}{B}$, from which we can read the slope $m = -\frac{A}{B}$ and the y-intercept $b = -\frac{C}{B}$

6. A horizontal line has slope = 0.
A vertical lines has an undefined slope.

7. Parallel Lines have the same slope.
Perpendicular Lines have slopes that are negative reciprocals, $m_1 = \frac{2}{3}$, $m_2 = -\frac{3}{2}$.
Then line 1 would be perpendicular to line 2.

WHAT YOU SHOULD BE ABLE TO DO upon completion of this unit.

1. Determine the slope of the line passing through two known points. Positive slope implies increasing function.

2. Write the equation of a line if given the slope and one point on the line, or if given two points on the line.

3. Write the equation of a line if given the slope and y-intercept of the line.

4. From the equation of a line find the slope and y-intercept.

5. From their equations determine if two lines are parallel, perpendicular, or neither.

6. If given a word relationship between two variables, write a linear equation expressing the relationship.

SOLUTIONS to Exercise Set 3.4

2. $m = \frac{-3 - 3}{-1 - 2} = \frac{-6}{-3} = 2$ increasing

5. $m = \frac{1 - 2}{\frac{3}{2} - \frac{1}{2}} = \frac{-1}{1} = -1$ decreasing

8. $m\,\overline{AB} = \frac{4 - 2}{3 + 3} = \frac{2}{6} = \frac{1}{3}$

$m\,\overline{BC} = \frac{-2 - 4}{5 - 3} = \frac{-6}{2} = -3$

$m\,\overline{CD} = \frac{-4 + 2}{-1 - 5} = \frac{-2}{-6} = \frac{1}{3}$

$m\,\overline{DA} = \frac{2 + 4}{-3 + 1} = \frac{6}{-2} = -3$

OPPOSITE SIDES Do have the same slopes, so we have a parallelogram.

11. Point (0, 0) tells us the intercept is 0, and using $y = mx + b$, we have $y = 3x$.

14. $P_1(-3, 5)$; $P_2(1, 7)$ $m = \frac{7 - 5}{1 + 3} = \frac{2}{4} = \frac{1}{2}$

Using $y - y_1 = m(x - x_1)$ we obtain

$y - 5 = +\frac{1}{2}(x + 3)$ or $y = +\frac{1}{2}x + 13$

17. $P_1\left(-\frac{1}{2}, -1\right)$; $P_2\left(\frac{1}{2}, -1\right)$ $m = \frac{-1 + 1}{\frac{1}{2} + \frac{1}{2}} = 0$ Horizontal line. Equation: $y = -1$

20. $m = -3$, $b = -3$ Use $y = mx + b$ to get

$y = -3x - 3$

23. $m = \frac{1}{3}$, $b = -5$ similar process: $y = \frac{1}{3}x - 5$

26. $2x - 5y + 3 = 0$ First solve for y.

$-5y = -2x - 3$

$y = \frac{2}{5}x + \frac{3}{5}$

From which $m = \frac{2}{5}$, $b = \frac{3}{5}$

29. $3x + 4y + 2 = 0$ $\quad m = -\frac{A}{B} = -\frac{3}{4}$

$b = -\frac{C}{B} = -\frac{1}{2}$

32. (a) Horizontal Line is of the form $y = k$; in this case $y = -2$.

(b) Vertical Line is of the form $x = k$; $x = -5$.

35. Similar to Exercise 32: (a) $y = -9$; (b) $x = 9$.

38. (a) Rearrange to read $-5x + 2y + 4 = 0$. Then the slope $= -\frac{A}{B} = \frac{5}{2}$ and this would also be the slope of all lines parallel to the given line.

(b) Perpendicular Lines have their slope = the negative reciprocal, or $-\frac{2}{5}$.

41. $(1, 3)$; $y = -3x + 2$ $\quad m = -3$

(a) Required line: $y - 3 = -3(x - 1)$ or $\underline{y = -3x + 6}$

(b) Perpendicular line: $y - 3 = \frac{1}{3}(x - 1)$ or $\underline{y = \frac{1}{3}x + \frac{8}{3}}$

44. $(-1, -3)$ $\quad 3y + 4x - 5 = 0$ $\quad m = -\frac{4}{3}$

(a) Parallel line: $y + 3 = -\frac{4}{3}(x + 1)$, $\underline{4x + 3y + 13 = 0}$

(b) Perpendicular line: $y + 3 = \frac{3}{4}(x + 1)$, $\underline{-3x + 4y + 9 = 0}$

47. (a) Essentially we have two points or pairs of values $P_1(10, 13.50)$ and $P_2(12, 15.90)$

$$m = \frac{15.90 - 13.50}{12 - 10} = \frac{2.40}{2} = 1.2$$

Now use $y - y_1 = m(x - x_1)$

$y - 13.50 = 1.2(x - 10)$

$y - 13.50 = 1.2x - 12$ or $\underline{y = 1.2x + 1.5}$

(b) Using this, if $y = 22$ we have

$22 = 1.2x + 1.5$ or $1.2x = 20.5$ and

$x = \$17.08$ approximately.

50. Given $cx - 5y + 8 = 0$ and that the x-intercept is 4 means that the line passes through (4, 0), which implies that $C \cdot 4 - 5 \cdot 0 + 8 = 0$ or $4C = -8$ and $C = -2$.

53. Given $f(x) = ax + b$ and $a > 0$, for $x_2 > x_1$, $ax_2 > ax_1$ and likewise regardless of b, $ax_2 + b > ax_1 + b$, which now means that $f(x_2) > f(x_1)$ whenever $x_2 > x_1$ and thus when $a > 0$ we have $f(x) = ax + b$ increasing.

A similar line of thought will prove the second part of the problem.

56. (a) $\angle CAQ = \angle BQC$ — Both are complements of $\angle CQA$.

(b) $\Delta ACQ \sim \Delta BCQ$ — Both are right triangles with corresponding angles equal.

(c) $\frac{\overline{CQ}}{\overline{AC}} = \frac{\overline{CB}}{\overline{CQ}}$ — Corresponding parts of similar triangles are proportional.

(d) $m_1 = \frac{\overline{CQ}}{\overline{AC}}$; $m_2 = -\frac{\overline{CQ}}{\overline{CB}}$ — Definition of slope $= \frac{y_2 - y_1}{x_2 - x_1}$

(e) $m_2 = -\frac{1}{m_1}$ — Steps (c) and (d).

Section 3.5 Direct and Inverse Variation

BASIC CONCEPTS you should learn from this unit.

1. If two variables vary DIRECTLY, the RATIO of the variables is constant. For example, $x/y = k$.

2. If two variables vary INVERSELY, the PRODUCT of the variables is constant. For example, $xy = k$.

3. If a variable varies JOINTLY as two or more other variables, it varies DIRECTLY as the PRODUCT of the other variables. For example, M varies jointly as P and q means

$$\frac{M}{P \cdot q} \longleftarrow \text{Product of P and q}$$

WHAT YOU SHOULD BE ABLE TO DO upon completion of this unit.

1. Write a relationship of Direct Variation, of Inverse Variation, of Joint Variation.

2. From appropriate data, determine a constant of variation.

3. Knowing the constant of variation and values for all but one of the variables, find the value of that variable.

SOLUTIONS for Exercise Set 3.5

2. (a) Inverse variation implies: $x \cdot y = k$, and for any x, y pair we see that $k = \underline{6}$.

(b) $y = \frac{6}{x}$ or $xy = 6$

(c)

x	...	12	15	18	24	60
y	...	$\frac{1}{2}$	$\frac{2}{5}$	$\frac{1}{3}$	$\frac{1}{4}$	$\frac{1}{10}$

5. (a) $\frac{s}{t^2} = k$ and $\frac{10}{100} = \frac{1}{10} = k$

(b) $s = kt^2$ therefore, $s = \frac{1}{10} \cdot 5^2 = 2.5$

8. (a) $V \cdot P^2 = k$: $\frac{2}{3} \cdot 6^2 = k$; $k = 24$

(b) $V = \frac{k}{P^2}$; $V = \frac{24}{8^2} = \frac{24}{64} = \frac{3}{8}$

11. (a) $\frac{Ms^2}{r^2} = k$; $\frac{4 \cdot 2^2}{4^2} = k = 1$

(b) If $r = 6$, $s = 5$; $\frac{M \cdot 25}{36} = 1$; $M = \frac{36}{25}$

14. (a) $\frac{A \cdot d^3}{b^2 \cdot c^2} = k$; $\frac{18 \cdot 8}{4^2 \cdot 3^2} = \frac{18 \cdot 8}{16 \cdot 9} = 1 = k$

(b) $A = \frac{kb^2c^2}{d^3} = \frac{1 \cdot 9^2 \cdot 4^2}{3^3} = \frac{16}{3}$

17. $R \cdot A = k$; $20 \cdot 8 = k = 160$

$R = \frac{K}{A} = \frac{160}{12} = \frac{40}{3}$ ohms

20. w = weight of body

s = distance

and $w \cdot s^2 = k$; $400 \cdot 4000^2 - k = 6,400,000,000$

$$w = \frac{k}{s^2} = \frac{64 \times 10^8}{(10^3)^2} = \frac{64 \times 10^8}{10^6} = 64 \times 10^2 = 6400 \text{ lb}$$

(1000 miles from the center would be 3000 miles below the surface.)

23. Let C = illumination

I = intensity

s = distance from source

$\frac{C \cdot s^2}{I} = k$ We have $\frac{50 \cdot 2^2}{400} = k$; $k = \frac{200}{400} \cdot \frac{1}{2}$

$$C = \frac{KI}{s^2} = \frac{\frac{1}{2} \cdot 3840}{4^2} = \frac{1920}{4^2} = 120 \text{ candlepower}$$

Section 3.6 Combining Functions; Inverse Functions

BASIC CONCEPTS you should learn from this unit.

1. Two or more FUNCTIONS can be combined by addition, subtraction, multiplication, and division with due care not to divide by zero.

2. A COMPOSITE FUNCTION is essentially a function of a function. In other words, we can apply one function to a variable and another function to the result. We often use the phrase "f of g of x" and write $(f \circ g)(x)$.

3. A ONE-TO-ONE FUNCTION is one in which for each x there is exactly one y and for each y there is exactly one x.

   Test: If a HORIZONTAL line intersects the graph of a function exactly once, then the function is one-to-one.

4. The inverse of a function is found by interchanging the roles of x and y and then solving for y.

   The inverse of a one-to-one function is also a function.

   $f^{-1}(x)$ is the notation for an inverse function.

5. Since y's become x's and vice versa in inverses, the inverse of a function is a reflection about the line $y = x$.

WHAT YOU SHOULD KNOW HOW TO DO upon completion of this unit.

1. Apply the four arithmetic operations to two or more functions.
2. Find Composite Functions.
3. Recognize one-to-one functions.
4. Use the horizontal line test.
5. Find the inverse of a function.
6. Show that $f\left[f^{-1}(x)\right] = x$.

SOLUTIONS to Exercise Set 3.6

2. $f(x) + g(x) = x^2 + 1 + x - 2$

   Therefore $(f + g)(2) = 2^2 + 1 + 2 - 2 = 5$

5. $f \cdot g(x) = f(x) \cdot g(x) = (x^2 + 1)(x - 2)$

   Therefore, $f \cdot g(2) = (2^2 + 1)(2 - 2) = 0$

8. $\frac{f}{g}(x) = \frac{f(x)}{g(x)} = \frac{x^2 + 1}{x - 2}$

   Therefore, $\left(\frac{f}{g}\right)(-2) = \frac{(-2)^2 + 1}{-2 - 2} = \frac{5}{-4}$

11. $f(x) = 2x + 1 \qquad g(x) = 2x^2 + x$

    $f \circ g = f[g(x)] = f(2x^2 + x) = 2(2x^2 + x) + 1 = 4x^2 + 2x + 1$

14. $(g \circ f)(3) = g[f(3)] = g(7) = 98 + 7 = 105$

17. $(g \circ f)(x - 1) = g[f(x - 1)] = g[2(x - 1) + 1] = g[2x - 1] = 2(2x - 1)^2 + (2x - 1)$

    $= 8x^2 - 6x + 1$

20. $f(x) = x^2 + 4 \qquad g(x) = \sqrt{x + 2}$

    $g[f(x)] = g(x^2 + 4) = \sqrt{x^2 + 4 + 2} = \sqrt{x^2 + 6}$

23. See solution 20. The domain of $(g\circ f)(x)$ is <u>all reals</u>.

26. $f(x) = \sqrt{x+1}$, $g(x) = x + 2$

(a) $(f\circ g)(x) = \sqrt{x + 2 + 1} = \sqrt{x + 3}$

(b) $(g\circ f)(x) = \sqrt{x + 1} + 2$

29. $f(x) = 2x + 4$, $g(x) = \frac{1}{2}x - 2$

$(f\circ g)(x) = 2\left(\frac{1}{2}x - 2\right) + 4 = x$

and

$(g\circ f)(x) = \frac{1}{2}(2x + 4) - 2 = x$

32. $f(x) = x^3$, $g(x) = \sqrt[3]{x}$

$(f\circ g)(x) = \left(\sqrt[3]{x}\right)^3 = x$

$(g\circ f)(x) = \sqrt[3]{x^3} = x$

35. $f(x) = 2x + 3$

$y = 2x + 3$

$x = 2y + 3$

$\frac{x - 3}{2} = y$

$f^{-1}(y) = \frac{x - 3}{2}$

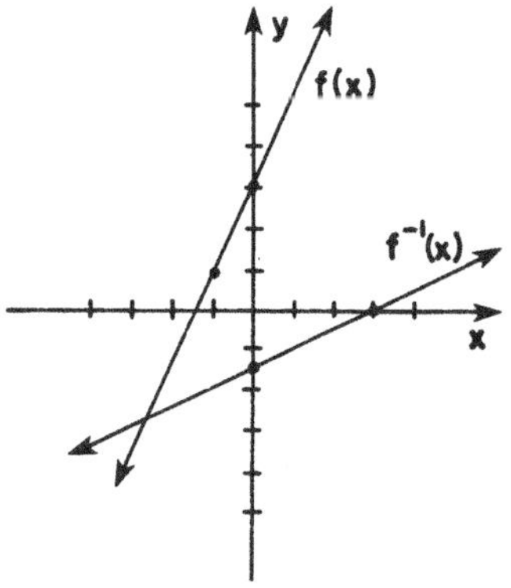

38. $f(x) = \frac{1}{2}x + 1$

$y = \frac{1}{2}x + 1$

Solving for x we have $x = 2y - 2$

Switch x and y. $f^{-1}(x) = y = 2x - 2$

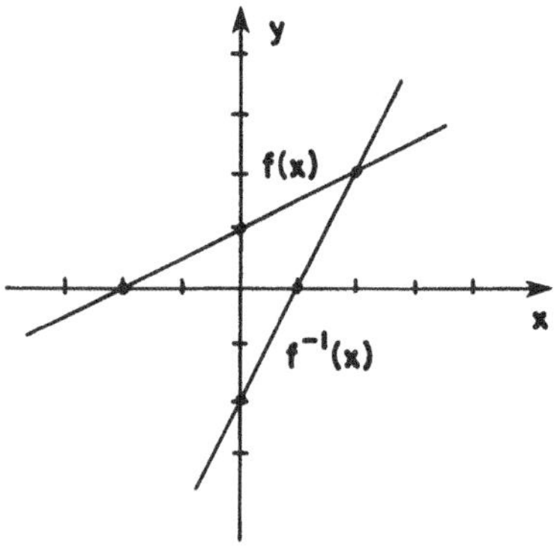

41. $f(x) = x^3 + 1$

$y = x^3 + 1$

$x = \sqrt[3]{y - 1}$

$f^{-1}(x) = y = \sqrt[3]{x - 1}$

x	0	1	2	9	-7
4	-1	0	1	2	-2

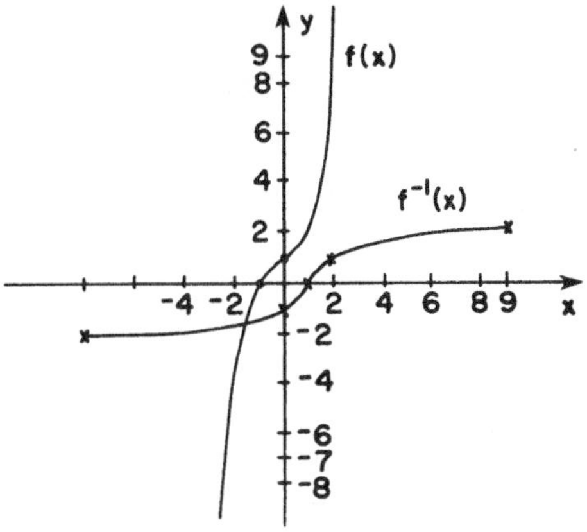

44. $f(x) = 3 - 5x$

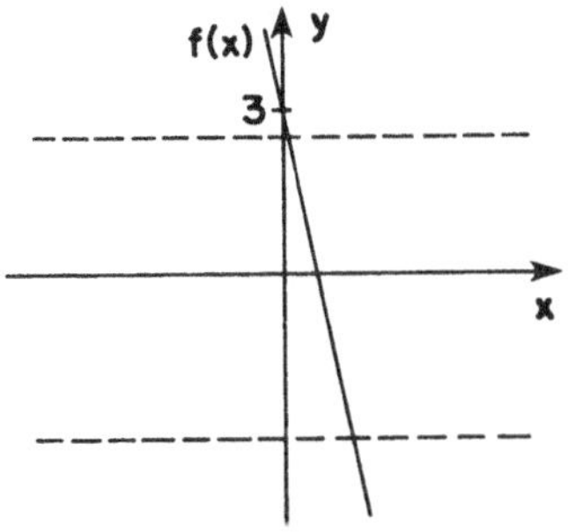

Yes, one-to-one.

47. $f(x) = -x^3 + 1$

x	0	1	2	-2
y	1	0	7	10

Yes, one-to-one

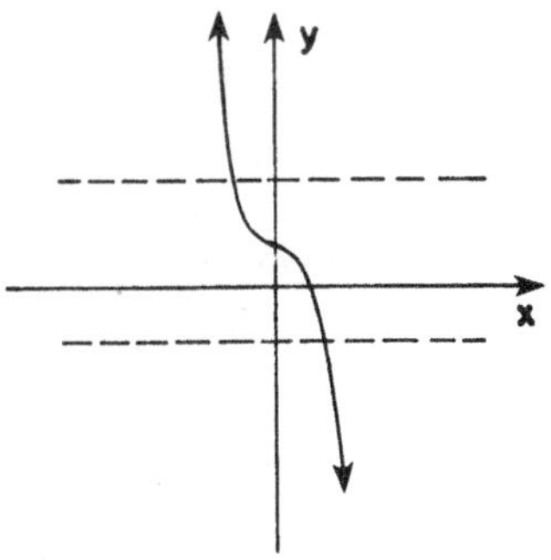

50. $f(x) = \begin{cases} x^2 - 4x + 4 & x \le 2 \\ x & x > 2 \end{cases}$

x	0	1	2	4
y	4	1	0	4

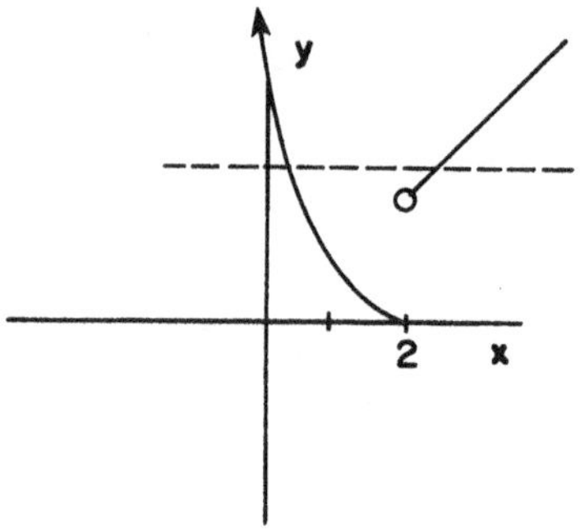

no, not one-to-one

53. $f(x) = ax + b$

$y = ax + b$

$x = \frac{y - b}{a}$

$y = \frac{x - b}{a} = f^{-1}(x)$

SOLUTIONS to Review Exercises for Chapter 3.

2. A(3, -4), B(-2, -6), C(-1, 2)

$\overline{AB} = \sqrt{(-2)^2 + 5^2} = \sqrt{29}$

$\overline{AC} = \sqrt{6^2 + 4^2} = \sqrt{52}$

$\overline{BC} = \sqrt{8^2 + 1^2} = \sqrt{65}$ Therefore, BC is the longest side.

5. $y^2 = 1 - x^3$ $-x$ for x means that $y^2 = 1 + x^3$ and changes the equation.
$-y$ for y implies $y^2 = 1 - x^3$.
There is no change; therefore there is x-axis symmetry only.

8. The vertical line test fails $\frac{1}{m}$ there is no function.

11. $f(x) = \sqrt{x - 1} = 15$ $15^2 + 1 = 226$
Thus if $\underline{x = 226}$, then $f(x) = \sqrt{226 - 1} = 15$.

14. $f(x) = x^2 - x$; $f(y - 1) = (y - 1)^2 - (y - 1) = \underline{y^2 - 3y + 2}$

17. $f(x) = \begin{cases} x - 1 & \text{for } x \leq -1 \\ x^2 & \text{for } -1 < x \leq 2 \\ -2 & \text{for } x > 2 \end{cases}$

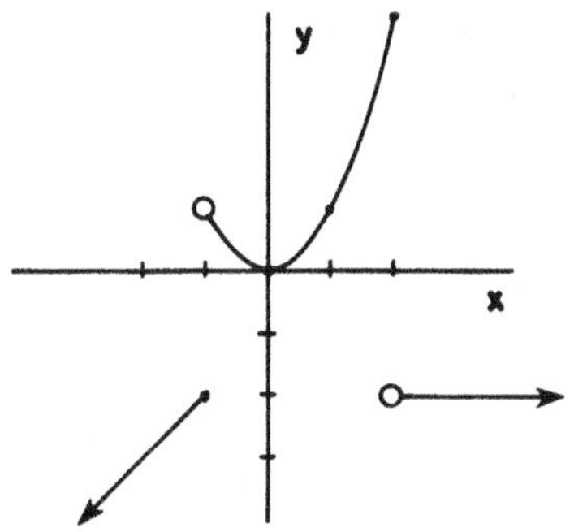

$f(x) \uparrow$ for $(-\infty, -1]$ and $(0, 2]$

$f(x) \downarrow$ for $(-1, 0)$

$f(x)$ constant for $x > 2$

20. A(-4, -6), B(-1, 3)

$m_{AB} = \frac{3 + 6}{-1 + 4} = \frac{9}{3} = 3$

23. Slope = 0, so $y = 0x + 3$ or $y = 3$.

26. Direct Variation implies that $R/q = k$. Therefore,

$20/5 = k = 4$. $R = kq = 4 \cdot 40 = \underline{160}$.

29. $f(x) = x + 1$ $g(x) = x^2 - 1$

$f(x) + g(x) = x + 1 + x^2 - 1 = x^2 + x$

32. $f(x) = x + 1$ $g(x) = x^2 - 1$

$\left(\frac{f}{g}\right) x = \frac{x + 1}{x^2 - 1}$ DOMAIN: All Reals except ±1, which cause denominator to be zero.

35. $f(x) = \sqrt{x} - 2$ $g(x) = x^2$

$(f \circ g)(x) = \sqrt{x^2} - 2$

38. $(g \circ f)(2) = (\sqrt{2} - 2)^2 = 2 - 4\sqrt{2} + 4 = 6 - 4\sqrt{2}$

Chapter 3 - A SHORT QUIZ

1. The Rectangular Coordinate System consists of two ___?___ Number Lines, one called the ___?___ , and the other called the ___?___ .
2. A point with abscissa of -2 would be in the ___?___ or ___?___ quadrant.
3. State the Distance Formula.
4. GRAPH the equation $y = |x - 2|$.
5. What kind of Symmetry does the graph of $y = x^2 + 1$ have?
6. If a rule assigns exactly one y-value to each x-value, the RULE is called a ___?___ .

7. What is the DOMAIN of $f(x) = \dfrac{x}{\sqrt{x - 1}}$?

8. Which of these are functions?

(a)

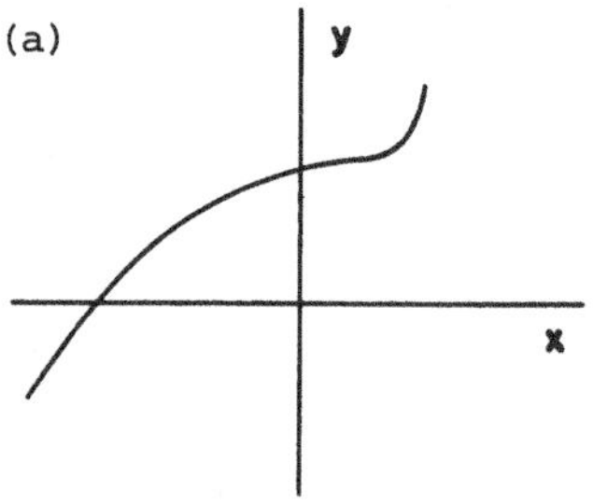

(b)

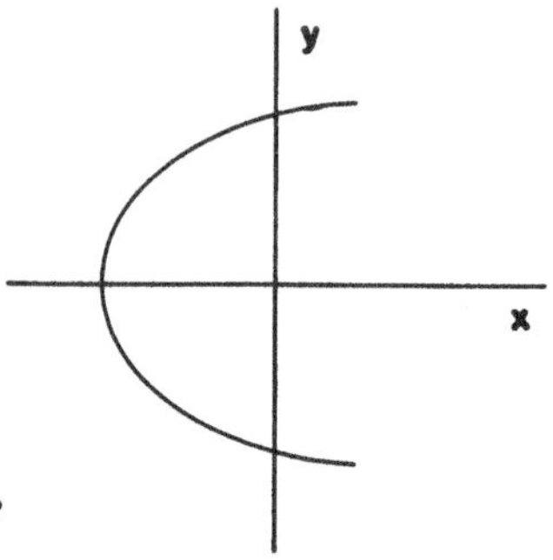

(c)

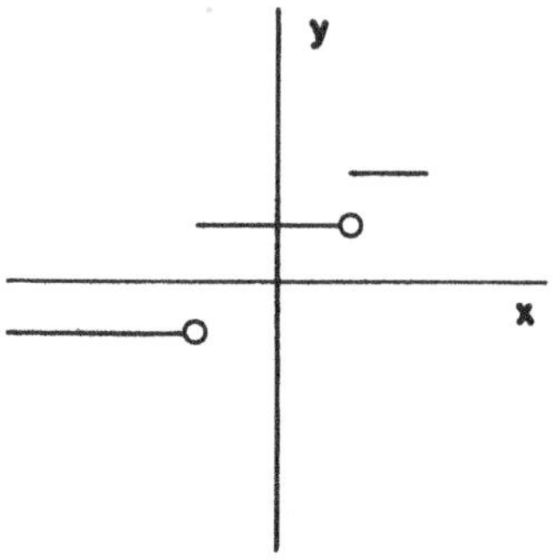

9. If $f(x) = 3x - 7$, what is $f(-2)$?

10. In Question 9 what is $f(a + h)$?

11. Given that $f(x_1) > f(x_2)$ whenever $x_1 < x_2$, is $f(x)$ increasing or decreasing?

12. By graphing $f(x) = x^2 - 2x + 1$ determine the interval where $f(x)$ is decreasing.

13. Is $f(x) = 7$ increasing, decreasing, or neither?

14. A function of the form $f(x) = mx + b$ is a ___?___ function.

15. Refer to Question 14: m represents the ___?___ of the line.

16. Generally, Slope is defined as equal to ___?___ .

17. Find the slope of a line passing through the points $P_1(3, -2)$; $P_2(-4, -7)$.

18. The graph of a decreasing linear function would have a ___?___ slope.

19. Write the point-slope form of the straight line equation.

20. If $2x - 5y - 7 = 0$, what is the slope? What is the y-intercept?

21. What kind of slope does a vertical line have?

22. If two lines are parallel, their slopes would be ___?___ .

23. Given that $x \cdot y = k$ we know that x and y vary ___?___ (directly or inversely).

24. Write an equation showing that x varies directly as the square of y and inversely as the square root of z .

In Questions 25 - 30 use the following: Given $f(x) = x^2 - 1$ and $g(x) = 2x$.

25. Find $(f + g)(x)$.

26. Find $\left(\frac{g}{f}\right)(x)$.

27. Find $(f \circ g)(x)$.

28. Find $f^{-1}(x)$.

29. Is $f^{-1}(x)$ a FUNCTION?

30. Find $g^{-1}(x)$.

Quiz Answers Chapter 3

1. perpendicular, x-axis, y-axis

2. second, third

3. $d = \sqrt{(x_2 - x_1)^2 + (y_2 - y_1)^2}$

4. $y = |x - 2|$

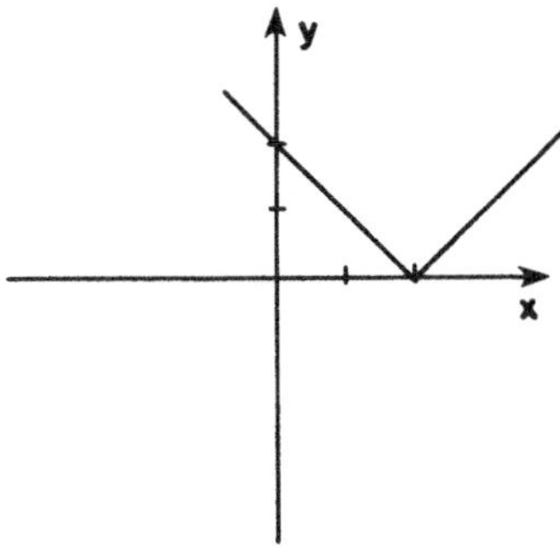

5. y-axis symmetry

6. function

7. D: $x > 1$

8. a, c

9. -13

10. $3a + 3h - 7$

11. decreasing

12. $(-\infty, 1)$

13. neither

14. linear

15. slope

16. $m = \dfrac{y_2 - y_1}{x_2 - x_1}$

17. $m = \frac{5}{7}$

18. negative

19. $y - y_1 = m(x - x_1)$

20. $\frac{2}{5}$, $-\frac{7}{5}$

21. undefined

22. equal

23. inversely

24. $\dfrac{x\sqrt{z}}{y^2} = k$

25. $x^2 + 2x - 1$

26. $\dfrac{2x}{x^2 - 1}$

27. $(2x)^2 - 1$

28. $f^{-1}(x) = \pm\sqrt{x + 1}$

29. no

30. $g^{-1}(x) = \frac{x}{2}$

CHAPTER 4

Exponential and Logarithmic Functions

Section 4.1 Exponential Functions

BASIC CONCEPTS you should learn from this unit.

1. Functions of the form $f(x) = a^x$ are called Exponential Functions. The variable, x, appears as an exponent on some base a.

   But there is a restriction: The base, a, must be positive $(a > 0)$ and not equal to one $(a \neq 1)$.

2. Some characteristics of exponential functions are

   A. $f(x) = a^x$ is increasing if $a > 1$;

      $f(x) = a^x$ is decreasing if $0 < a < 1$.

   B. $f(x) = a^{-x}$ is decreasing if $a > 1$.

      $f(x) = a^{-x}$ is increasing if $0 < a < 1$.

   C. If $a^x = a^y$, then $x = y$ and if $a^x = b^x$ then $a = b$.

3. Three applied examples are discussed in the text:

   A. The Growth Model: $Q(t) = q_0e^{kt}$. Here time is the usual variable and since the base, e, equals approximately 2.71828, which is greater than 1, the function is increasing (growth).

   B. The Decay Model: $Q(t) = q_0e^{-kt}$. This is of the form $f(x) = a^{-x}$ (see concept 2B) and so is decreasing, since the base, again e, is greater than 1.

   C. The Compound Interest Formula

      i. $S = P(1 + i)^n$ Here the base $(1 + i)$ is always greater than 1, so we have an increasing function.

      ii. $S = Pe^{rt}$ for continuous compounding.

WHAT YOU SHOULD KNOW HOW TO DO upon completion of this unit.

1. Graph exponential functions with various bases and exponents.

2. Solve simple exponential equations such as $2^x = 2^3$ and $(a + 1)^x = (2a - 1)^x$.

3. Use the growth model, decay model, and the compound interest formulas to solve applied problems.

SOLUTIONS to Exercise set 4.1

1. $f(x) = 4^x$

   $f(0) = 1$; $f(2) = 16$

   $f(-2) = 1/16$

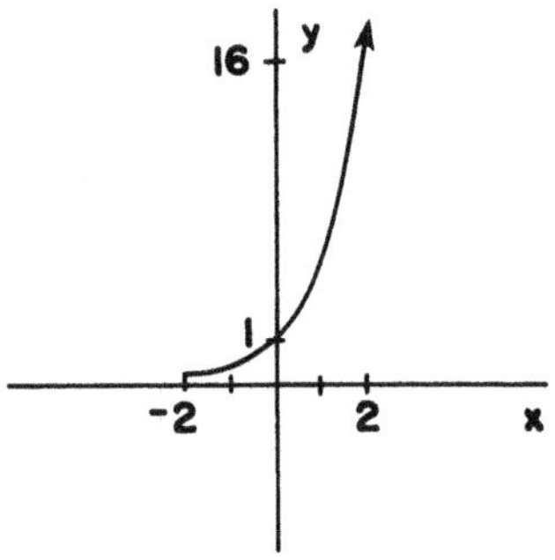

4. $f(x) = 10^{-x}$

$f(0) = 1; \quad f(1) = \frac{1}{10}$

$f(-1) = 10; \; f(-2) = 100$

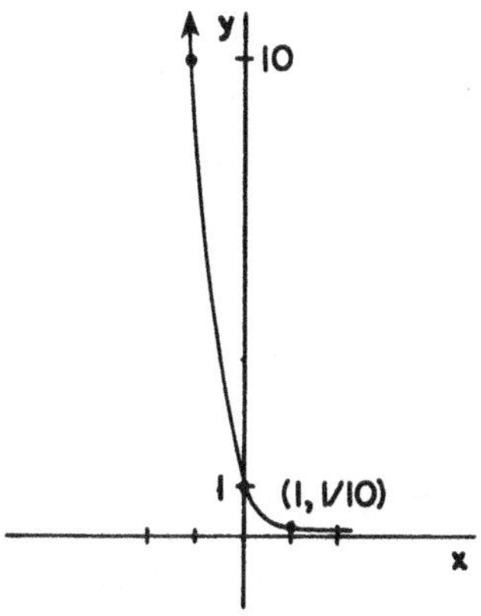

6. $f(x) = 2^{x-1}$

$f(0) = 2^{-1} = \frac{1}{2}$

$f(1) = 2^{0} = 1$

$f(2) = 2^{1} = 2$

$f(-1) = 2^{-2} = \frac{1}{4}$

Note that this graph of $f(x) = x^{x-1}$ is just the graph of $f(x) = 2^{x}$ shifted to the right one unit point for point.

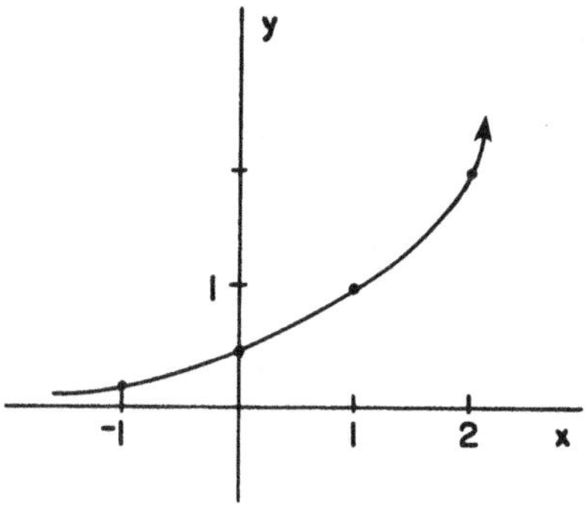

7. $f(x) = 2^{|x|}$

Here the exponent on 2 is always non negative.

$f(0) = 1; \quad f(1) = 2$

$f(-1) = 2; \quad f(2) = 4$

$f(-2) = 4$

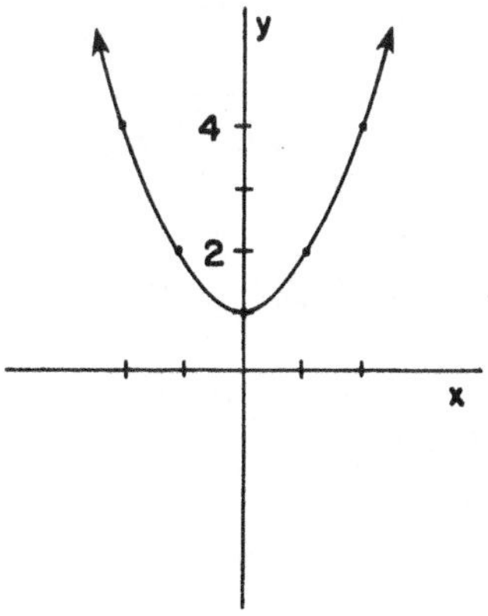

10. $f(x) = 3^{-2x}$

The exponent is opposite in sign to x, so $f(x)$ is decreasing.

$f(0) = 1$ $\quad f(-1) = 3^2$

$f(1) = 1/3^2$ $\quad f(-2) = 3^4$

$f(2) = 1/3^4$

The graph is decreasing very rapidly when $x < 0$ and decreases very slowly after $x > 0$.

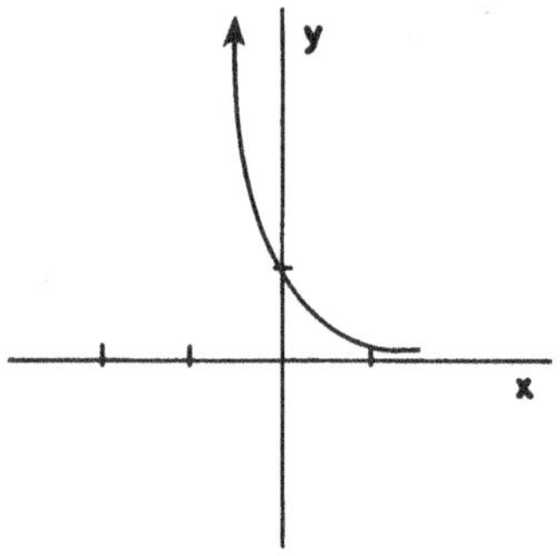

15. $3^x = 9^{x-2}$

$3^x = (3^2)^{x-2} = 3^{2x-4}$ using the $(x^m)^m = x^{mm}$ property of exponents

Thus $x = 2x - 4$

and $x = 4$

Check: If $3^4 = 9^{4-2}$, then $3^4 = 9^2$

$81 = 81$

19. $e^{x-1} = e^3$

$x - 1 = 3$

$x = 4$

20. $e^{x-1} = 1$

$x - 1 = 0$ Because $e^0 = 1$

Thus $x = 1$

23. $(a + 1)^x = (2a)^x$

$a + 1 = 2a$

$1 = a$

25. a. The basic formula to use here is the exponential growth model $Q(t) = q_0e^{kt}$. Thus we see from $Q(t) = 200e^{0.25t}$ that $q_0 = 200$; this is the initial number of bacteria. Also, $k = 0.25$.

b. $t = 20$, so $Q(t) 200 \cdot e^{(0.25)(20)} = 200e^5 = 29683$

c.

t	1	4	8	10
Q	257	544	1478	2436

In the basic formula $Q(t) = 200e^{(0.25)5}$, let $t = 1, 4, 8, 10$ successively to obtain the Q values in the table.

29. Using the decay formula, $Q(t) = q_0e^{-kt}$ with $q_0 = 1000$, we obtain, when the rate of decay is 4%, $k = 0.04$, and $t = 10$. Thus $Q(t) = 1000_e{}^{(0.04) \cdot 10} = 1492$ grams.

32. We use the compound interest formula $S = P(1 + i)^n$, in which n = number of conversion periods and $i = r/k$, r being the annual interest rate, and k is the number of conversion periods per year.

For Investment A $r = 8\%$ and $k = 4$, and so

$S = P(1 + 0.02)^4 = P(1.0824)$

For Investment B $r = 9\%$ and $k = 1$, and so

$S = P(1.09)^1 = P(1.09)$

35. Using the continuous compounding formula $S = Pe^{rt}$ with $P = 5000$, $r = 0.09$, and $t = 5$, we have

$S = 5000e^{(0.09)(5)} = 5000e^{0.45} = 5000(1.5683)$

therefore

$S \approx \$7842$

39. INVESTMENT A:

$S = 1000(1.04)^{20}$

$S = 2191.12$

INVESTMENT B:

$S = 1000e^{0.08 \cdot 10} = 1000e^{0.8}$

$S = 2225.54$

The approximate difference is thus $34.42.

Section 4.2 Logarithmic Functions

BASIC CONCEPTS you should learn from this unit.

1. The logarithmic function $y = \log_a x$ is the inverse of the exponential function $y = a^x$.
2. The domain of the logarithmic function is restricted: $x > 0$.
3. The base, a, of the logarithmic function is restricted: $a > 0$ and $a \neq 1$. (Note that this is exactly the same restriction the base, a, has in the exponential function $y = a^x$.)
4. The statement of the logarithmic function, $y = \log_a x$ if and only if $x = a^y$, tells us that if we raise a to the power y we get x; in other words, y is an exponent.
5. $\ln x = \log_e x$ is the usual way to write the base e logarithmic function.
6. $a^{\log_a x} = x$ and $\log_a a^x = x$ are two simple identities to be learned.

WHAT YOU SHOULD KNOW HOW TO DO upon completion of this unit.

1. Change back and forth from the exponential form to the logarithmic form; for example,

$$3 = \log_2 8 \text{ if and only if } 8 = 2^3$$

2. Solve elementary log equations; for example,

$\log_2(x - 1) = \log_2 10$

means

$x - 1 = 10$

so

$x = 11$

3. Use the identities and concepts above to evaluate expressions; for example,

$2^{\log_2 7} = 7$

$\log 0.001 = -3 \quad (10^{-3} = 0.001)$

4. Graph simple log functions.

SOLUTIONS to Exercise Set 4.2

In many of the exercises we use the basic concept that the logarithm of a number to some base is the exponent we would use on the base to obtain the number (see Basic Concept 4). Thus:

1. $\log_2 4 = 2$ means $2^2 = 4$

3. $\log_9 \frac{1}{81} = -2$ means $9^{-2} = \frac{1}{81}$

5. $\ln 20.09 = 3$ means $e^3 = 20.09$

9. $\ln 1 = 0$, $e^0 = 1$

13. $25 = 5^2$ implies $\log_5 25 = 2$

17. $\frac{1}{8} = 2^{-3}$ implies $\log_2 \frac{1}{8} = -3$

21. $1 = e^0$ implies $\log_e 1 = 0$ or $\ln 1 = 0$

25. $\frac{1}{3} = 27^{-1/3}$ implies $\log_{27} \frac{1}{3} = -\frac{1}{3}$

29. $\log_{25} x = \frac{1}{2}$ implies $25^{1/2} = x$, so $x = 5$, $x = \sqrt{25}$. Recall that $x > 0$.

33. $\ln x = \frac{1}{2}$, $e^{1/2} = x$, $x = \sqrt{e} \approx 1.649$

37. $\log_x \frac{1}{8} = \frac{1}{3}$, $x^{1/3} = \frac{1}{8}$, then cubing both sides gives $x = \frac{1}{512}$

41. $\log_{x+1} 24 = \log_3 24$ implies that $x + 1 = 3$, $x = 2$

45. $3^{\log_3 6} = 6$

49. $\log_5 5^3 = 3$

53. $\log_7 49 = \log_7 7^2 = 2$

57. $\ln 1 = 0$, $(e^0 = 1)$

61. $\log 10{,}000 = 4$, $(10^4 = 10{,}000)$

65. $f(x) = \log_4 x$

x	1	$\frac{1}{4}$	4	16
y	0	-1	1	2

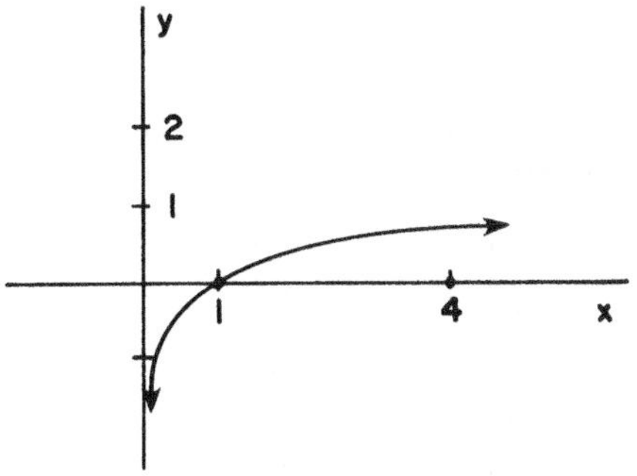

69. $f(x) = \ln \frac{x}{2}$ Here the base is e, so the function values (y-values) will be powers of e, yielding x/2 .

$\ln \frac{x}{2} =$

x	2	1	4	6	8
y	0	-0.693	0.693	1.098	1.386

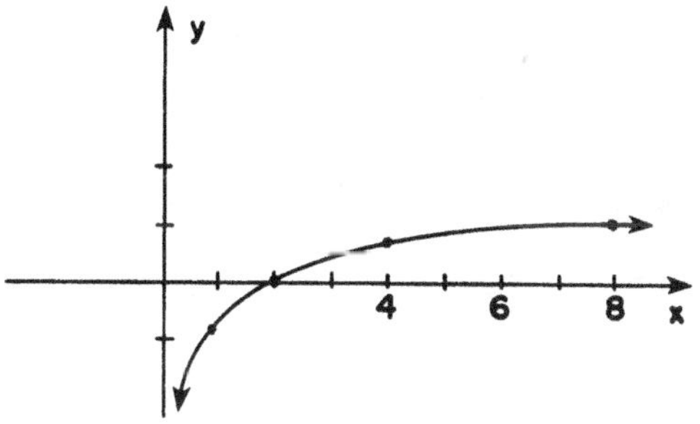

Section 4.3 Properties of Logarithms

BASIC CONCEPTS you should learn from this unit.

1. There are three properties of logarithms and they are virtually the same as the properties of exponents (which you already know). This is not too surprising, since logarithms really are exponents.

   A. $\log_a(x \cdot y) = \log_a x + \log_a y$ In words this equation says that
   The Log of a product equals the sum of the logs of the factors.

   B. $\log_a \left(\frac{x}{y}\right) = \log_a x - \log_a y$ In words this equation states that
   The Log of a Quotient equals the Log of the DIVIDEND MINUS the Log of the Divisor.

   C. $\log_a x^m = m \log_a x$ In words this equation says that
   The Log of a number raised to a power equals the exponent multiplied by the Log of the number.

That's all! But a couple of warnings should be mentioned:

A. Remember that $\log_a(x + y) \neq \log_a x + \log_a y$.

B. Remember that $\log_a(x - y) \neq \dfrac{\log x}{\log y}$.

If you have not yet memorized the three properties listed above, you should do so now! Solving problems will be much easier when you have learned them.

WHAT YOU SHOULD KNOW HOW TO DO upon completion of this unit.

1. Change logarithmic expression into simpler logarithmic form; for example,

$$\ln 2x^3 = \ln 2 + 3 \ln x$$

2. Write expressions containing several logarithms as a single logarithm; for example,

$$3 \log_a x - 2 \log_a z = \log_a \frac{x^3}{z^2}$$

3. Use the properties of logarithms to find the logarithms of products, quotients, roots, etc; for example, if $\log 2 = 0.30$ and $\log 3 = 0.47$, then

$\log 12 = \log 2 \cdot 2 \cdot 3 = \log 2 + \log 2 + \log 3$ and so

$\log 12 = 0.30 + 0.30 + 0.47 = 1.07$

SOLUTIONS to Exercise Set 4.3

1. $\log_{10}(120 \times 36) = \log_{10} 120 + \log_{10} 36$, using the Product Rule (Property 1.A).

3. $\text{Log}_3 3^4$ — Parentheses are not really necessary, are they?

$= 4 \log_3 3$ — Power Rule (Property 1.C)

But

$\log_3 3 = 1$

So

$4 \log_3 3 = 4$

7. $\log_a\left(\dfrac{x}{yz}\right) = \log_a x - [\log_a y + \log_a z]$

$= \log_a x - \log_a y - \log_a z$

11. $\log_a(x^2 y^3) = \log_a x^2 + \log_a y^3$

product rule

$= 2 \log_a x - 3 \log_a y$

power rule

13. $\log_a \sqrt{xy} = \log_a (xy)^{1/2} = \log_a x^{1/2} y^{1/2}$ by properties of Radicals and Exponents

then:

$\log_a x^{1/2} + \log_a y^{1/2}$

(by the product property), and finally

$\frac{1}{2} \log_a x + \frac{1}{2} \log_a y$

by the Power Rule

17. $\ln(\sqrt{x}\sqrt[3]{y}) = \ln x^{1/2} + \ln y^{1/3} = \frac{1}{2} \ln x + \frac{1}{3} \ln y$

21. $\log 6 = \log (2 \cdot 3)$ Think of 6 as a product.

so $= \log 2 + \log 3$

$= 0.30 + 0.47 = 0.77$

24. $\log \sqrt{5} = \log 5^{1/2} = \frac{1}{2} \log 5 = \frac{1}{2}(0.70) = 0.35$

27. $\log \frac{15}{2} = \log \frac{3 \cdot 5}{2} = \log 3 + \log 5 - \log 2$

$= 0.47 + 0.70 - 0.30$

$= 0.87$

29. $\log \sqrt{7.5} = \log(7.5)^{1/2} = \frac{1}{2} \log \frac{15}{2}$

$= \frac{1}{2}(0.87)$ (from Exercise 27)

$\log \sqrt{7.5} = 0.435$

33. $\frac{1}{3} \ln x + \frac{1}{3} \ln y = \ln x^{1/3} + \ln y^{1/3} = \ln x^{1/3} \cdot y^{1/3} = \ln(xy)^{1/3}$ <u>or</u> $\ln \sqrt[3]{xy}$

35. $\frac{1}{3} \log_a x + 2 \log_a y - \frac{3}{2} \log_a z = \log_a x^{1/3} + \log_a y^2 - \log_a z^{3/2} = \log_a \frac{x^{1/3} y^2}{z^{3/2}}$

37. $\frac{1}{2}(\log_a x + \log_a y) = \frac{1}{2} \log_a x + \frac{1}{2} \log_a y = \log_a x^{1/2} + \log_a y^{1/2} = \log_a x^{1/2} y^{1/2}$

$= \log_a (xy)^{1/2}$ or $\log_a \sqrt{xy}$

41. $\frac{1}{2} \log_a (x - 1) - 2 \log_a (x + 1) = \log_a \frac{(x - 1)^{1/2}}{(x + 1)^2}$ or $\log_a \frac{\sqrt{x - 1}}{(x + 1)^2}$

43. $3 \log_a x - 2 \log_a (x - 1) + \frac{1}{2} \log \sqrt[3]{x + 1} = \log_a x^3 - \log_a (x - 1)^2 + \log \left[(x + 1)^{1/3}\right]^{1/2}$

$= \log \frac{x^3 \cdot (x + 1)^{1/6}}{(x - 1)^2}$

BASIC CONCEPTS you should learn from this unit.

1. The logarithm of a number has two parts, a <u>mantissa</u>, which is always positive, and a <u>characteristic</u>, which is any integer, positive, negative, or zero.

2. Common logs are BASE 10 logs.
   Natural logs are BASE e logs.

3. Log Tables supply only the mantissa. You must determine the characteristic.

4. Computation with logs utilizes the laws of logarithms and requires a knowledge of the use of log tables.

WHAT YOU SHOULD KNOW HOW TO DO upon completion of this unit.

1. Change the form of a number to Scientific Notation.

2. Given a number, find its common logarithm and/or natural logarithm.

3. Given the logarithm of a number, find the number.

4. Do various computations using three-digit numbers by means of logarithms.

SOLUTIONS to Exercise Set 4.4

1. $2725 = 2.725 \times 10^3$

4. $0.000914 = 9.14 \times 10^{-4}$

7. $296.2 = 2.962 \times 10^2$

10. $\ln 3.2 = 1.1632$

11. $\log 37.5$: characteristic $= 1$
    mantissa $= 0.5740$
    answer $= 1.5740$

14. $\ln 60 = \ln(5 \cdot 12)$
    $= \ln 5 + \ln 12$
    $= 1.6094 + \ln 3 + \ln 4$
    $= 1.6094 + 1.0986 + 1.3863$
    $= \underline{4.0943}$

17. $\log 48{,}200$ = characteristic + mantissa
    $4 + 0.6830 = 4.6830$

20. $\log 0.00532 = -3 + 0.7259 = \underline{0.7259 - 3}$

23. $\ln x = 1.0647$. Then from Table II, $x = 2.90$.

26. $\log x = 0.1903 - 2$. Then from the table, $x = 1.55 \times 10^{-2}$ <u>or</u> $x = 0.0155$.

29. $\log x = 0.7832 - 4$. Then from the table, $x = 6.07 \times 10^{-4}$ <u>or</u> $x = 0.000607$.

32. log x = -3,9004. Convert first to a <u>positive</u> mantissa:

$$\begin{array}{r} +4.0000 \\ -3.9004 \\ \hline +0.0996 \end{array} \quad \begin{array}{l} \\ \text{So } -3.9004 \text{ is} \\ \text{equivalent to } 0.0996 - 4 \end{array}$$

This implies that $x = 1.26 \times 10^{-4}$, approximately, since 0.0996 is not in the table.

35. $N = \frac{679}{321}$

$$\begin{aligned} \log 679 &= 2.8319 \\ -\log 321 &= \underline{2.5065} \\ \log N &= 0.3254 \\ N &= 2.12 \times 10^{0} = 2.12, \text{ approximately.} \end{aligned}$$

38. $N = (42.3)^3(71.2)$

$$\begin{aligned} \log N &= 3 \log(42.3) + \log 71.2 \\ &= 3(1.6263) + 1.8525 \\ &= 4.8789 + 1.8525 \\ &= 6.7314 \end{aligned}$$

and $N = 5.39 \times 10^6 = 539{,}000$ approximately.

41. $N = \frac{\sqrt{7870}}{(46.3)^4}$

$$\begin{aligned} \log N &= \frac{1}{2} \log 7870 - 4(\log 46.3) \\ &= \frac{1}{2}(3.8960) - 4(1.6656) \\ &= 1.9480 - 6.6624 = -4.7144 \\ &= 0.2856 - 5 \end{aligned}$$

Then $N = 1.93 \times 10^{-5}$ or 0.0000193

44. $T = 2\pi\sqrt{\frac{L}{G}}$ Given $L = 4.72$, $g = 32.2$, $\pi = 3.14$; so

$$T = 2(3.14)\sqrt{\frac{4.72}{32.2}} = 6.28\sqrt{\frac{4.72}{32.2}}$$

$$\begin{aligned} \frac{1}{2}(\text{Log of radicand}) &= \frac{1}{2}(\log 4.72 - \log 32.2) \\ &= \frac{1}{2}(0.6739 - 1.5079) = \\ &= \frac{1}{2}(-0.8340) = -0.4170, \text{ or} \\ &= 0.5830 - 1 \end{aligned}$$

Now add $\log 6.28 = \underline{0.7980}$

so $\log T = 1.3810 - 1 = 0.3810$ (from table)

or $\underline{T = 2.404}$

47. The compound interest formula is

$$A = P(1 + r)^n = P(1 + \frac{r}{k})^n$$

In this problem, $n = 10$, $k = 2$, $r = 7.8\%$; so

$$A = 10{,}000\left(1 + \frac{0.078}{2}\right)^{10} = 10{,}000(1.039)^{10}$$

$$\log A = \log 10{,}000 + 10 \log 1.039$$

$$\log A = 4 + 0.1662 = 4.1662$$

Thus $A = \$14660$ Approximately

50. The formula is $A = \sqrt{s(s - a)(s - b)(s - c)}$ with

$s = \frac{1}{2}(a + b + c) = \frac{1}{2}(12.86 + 13.72 + 20.3) = 23.44$; $A = \sqrt{23.44(10.58)(9.72)(3.14)}$

Now using logs we have

$$\log A = \frac{1}{2}(\log 23.44 + \log 10.58 + \log 9.72 + \log 3.14)$$

$$= \frac{1}{2}(1.3692 + 1.0253 + 0.9877 + .4969)$$

$\log A = 1.9396$

and

$A = 87.0'$ approximately

Section 4.5 Exponential and Logarithmic Equations

BASIC CONCEPTS you should learn from this unit.

1. An Exponential equation is one in which the variable appears in an exponent (for example, $2^{2x-1} = 56$). To solve equations of this type we usually take the Logarithm of both sides and then apply the laws of logarithms and rules for solving equations.

2. A Logarithmic equation is one in which we find the log of an expression containing the variable; for example, $\log x + \log(x + 21) = 2$.

   Such equations are normally solved by converting the logarithmic expression to one logarithm, then changing to exponential form.

WHAT YOU SHOULD KNOW HOW TO DO upon completion of this unit.

1. Recognize and solve EXPONENTIAL EQUATIONS.

2. Recognize and solve LOGARITHMIC EQUATIONS.

SOLUTIONS to Exercise Set 4.5

2. $2^x = 24$; $\log 2^x = \log 24$; $x \log 2 = \log 24$

   Therefore, $x = \log 24/\log 2 = 1.3802/0.3010 = 4.585$

5. $3^{2x} = 46$ implies that $\log 3^{2x} = \log 46$ or $2x \log 3 = \log 46$

   Therefore, $2x = \log 46/\log 3 = 1.6628/0.4771 = 3.4852$; $x = 1.743$

8. $3^{3x-2} = 23.1$ implies that $(3x - 2)\log 3 = \log 23.1$ implies that
$3x - 2 = \log 23.1/\log 3 = 2.8580$: $3x = 4.8580$
$\underline{x = 1.619}$

11. $2^{-x} = 15$ implies that $-x(\log 2) = \log 15$ or $-x = 3.907$
$x = -3.907$

14. $3^{-3x+2} = 2^{-x}$ implies that $(-3x + 2)\log 3 = -x(\log 2)$ or
$(-3x + 2)(0.4771) = -0.3010x$; $-1.4313x + 0.9542 = -0.3010x$
$0.9542 = 1.1303x$; $x = 0.844$

17. $e^{2x+3} = 20$ implies that $(2x + 3)\ln e = \ln 20$ (natural log here because of e) implies that $2x + 3 = \ln 20 = 2.9957$

$2x = -0.0043$; $\underline{x = -0.0022}$ approximately

20. $\log x - \log 3 = 2$; $\log \frac{x}{3} = 2$; $\frac{x}{3} = 10^2 = 100$,
so $x = 300$.

23. $\log x + \log(x - 3) = 1$; $\log x(x - 3) = 1$; $x^2 - 3x = 10$
$x^2 - 3x - 10 = 0$ $(x - 5)(x + 2) = 0$; $x = 5$, $x = -2$
Discard the negative root; so $x = 5$.

26. $\log(7x - 2) - \log(x - 2) = 1$; $\log\left(\frac{7x - 2}{x - 2}\right) = 1$ implies that
$\frac{7x - 2}{x - 2} = 10$; $7x - 2 = 10x - 20$; $\underline{x = 6}$

29. $\log_2(x + 4) = 3 - \log_2(x - 2)$ implies that $\log_2(x + 4) + \log_2(x - 2) = 3$
$\log_2(x + 4)(x - 2) = 3$ implies that $x^2 + 2x - 8 = 8$; $x^2 + 2x - 16 = 0$
$x = 3.123$ approximately

32. $50{,}000 = 20{,}000\, e^{0.05t}$ implies that $2.5 = e^{0.05t}$
$\ln 2.5 = 0.05t \ln e = 0.05t$
$0.9163 = 0.05t$ so $t = \underline{18.33}$ years

35. Use $A = P\left(1 + \frac{r}{k}\right)^n$. Let $P = 1$ and $A = 2$.
Since we wish the amount to double,
$2 = \left(1 + \frac{.08}{2}\right)^n$; $2 = (1.04)^n$; $\log_2 = n \log 1.04$
$n = 0.3010/0.0170 = 17.7 = 18$. NOW n is the number of compounding periods and we are compounding semiannually, so it would take 9 years.

38. $430 = 500e^{3k}$ $\quad \frac{430}{500} = e^{3k}$; $\ln 430 - \ln 500 = 3k$

$3k = -0.1508$; $\underline{k = 0.05}$ = 5% rate of decay

SOLUTIONS to Review Exercises

2. $2^{2x} = 8^{x-1}$ Note that $8 = 2^3$ so $2^{2x} = (2^3)^{x-1}$.

$2^{2x} = 2^{3x-3}$ and $2x = 3x - 3$; $x = 3$

5. $27 = 9^{3/2}$ implies that $\log_9 27 = \frac{3}{2}$

6. $8 = 64^{1/2}$

10. $\log_5 \frac{1}{125} = x - 1$ implies that $\frac{1}{125} = 5^{x-1}$ and since $5^{-3} = \frac{1}{125}$

$5^{-3} = 5^{x-1}$ and $-3 = x - 1$; $\underline{x = -2}$

13. $\log_3 3^5 = 5$

16. $e^{\ln 3} = 3$

19. $\log_a \frac{x(2 - x)^2}{(y + 1)^{1/2}} = \log_a x + 2 \log_a (2 - x) - \frac{1}{2} \log_a (y + 1)$

22. $14 = 2 \cdot 7$ so $\log 14 = \log 2 + \log 7$

$= 0.30 + 0.85$

$\log 14 = 1.15$

25. $0.7 = 7 \times 10^{-1}$, so $\log 0.7 = 0.85 - 1$

28. $\ln 3x + 2(\ln y - \frac{1}{2} \ln x) = \ln 3x + 2 \ln y - \ln x = \ln \frac{3xy^2}{x} = \ln 3y^2$

31. $0.098 = 9.8 \times 10^{-2}$

34. $N = (0.765)(32.4)^2$; $\log N = \log 0.765 + 2 \log 32.4$

$= 0.8837 - 1 + 3.0211 = 2.9048$

$N = 803$ approximately

37. The Exponential Decay Model is $Q(t) = Q_0 e^{-kt}$ (see Section 4.1). If $Q_0 = 1$, we will then want $Q(t) = \frac{1}{2}$. $\frac{1}{2} = e^{-0.06t}$; $\ln \frac{1}{2} = \ln e^{-0.06t}$;

$\ln 1 - \ln 2 = -0.06t$; $\ln 2 = 0.06t$ implies that $t = 11.5$ hr.

40. $\log(2x - 1) = 2 + \log(x - 2)$; $\log(2x - 1) - \log(x - 2) = 2$;

$\log \frac{2x - 1}{x - 2} = 2$; $\frac{2x - 1}{x - 2} = 100$ implies that $2x - 1 = 100x - 200$

$x = \frac{199}{98}$

Chapter 4 - A SHORT QUIZ

1. Functions of the form $f(x) = a^x$ are called ____?____ functions.

2. For functions like $f(x) = a^x$ what restriction is placed on a?

3. The graph of the function $f(x) = a^x$ will be ____?____ (increasing or decreasing) if $a > 1$.

4. Graph the function $f(x) = 3^{-x}$.

5. If $b^5 = z^5$, what do we know about b and z?

6. Solve for x : $2^x = 8^{x+2}$

7. If the number of bacteria in a culture is described by $Q(t) = 80e^{0.3t}$, and t is measured in hours, (a) what is the initial number of bacteria in the culture? (b) Find the number of bacteria after 25 hours.

8. The inverse of the exponential function $y = 2^x$ is ____?____.

9. Write $\log_{3b} 6 = \frac{1}{2}$ in exponential form.

10. Solve for x : $\log_3 x = -4$.

11. Graph $y = \log_3 x$.

12. Solve for x : $\log_2 x^2 = \log_2 9$.

13. Complete these statements of the properties of logarithms.

    a. $\log_a(x \cdot y) =$ ____?____

    b. $\log_a\left(\frac{x}{y}\right) =$ ____?____

    c. $\log_a x^n =$ ____?____

14. Write $\ln(4xyz)$ in terms of simpler logs.

15. Write $\frac{2}{3}(4 \ln x - 3 \ln y)$ as a single logarithm.

16. Approximate $N = \dfrac{\sqrt[4]{47.9}}{(3.12)^3}$ using logarithms.

17. Solve for x : $5^{2x-3} = 19$

18. Solve for x : $\log(x + 197) - \log(x - 1) = 2$

Quiz Answers Chapter 4

1. Exponential

2. $a > 0$, $a \neq 1$

3. increasing

4.

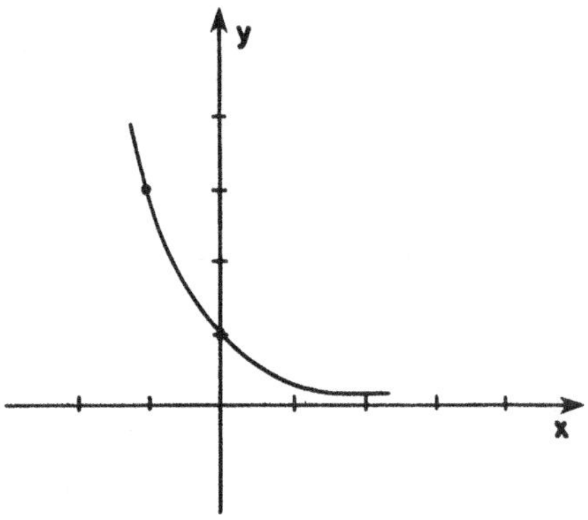

5. $b = z$

6. $x = -3$

7. a. 80

   b. 144, 644

8. $y = \log_2 x$

9. $36^{1/2} = 6$

10. $x = 1/81$

11. $y = \log_3 x$

x	y
1	0
3	1
$\frac{1}{3}$	-1

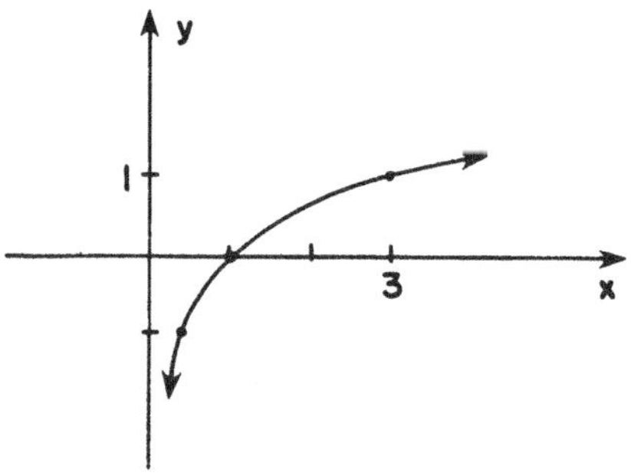

12. $x = \pm 3$

13. a. $\log_a x + \log_a y$

    b. $\log_a x - \log_a y$

    c. $n \log_a x$

14. $\ln 4 + \ln x + \ln y + \ln z$

15. $\ln \frac{x^{8/3}}{y^2}$

16. 0.0866

17. $x = 2.415$

18. $x = 3$

CHAPTER 5

Analytic Geometry: The Conic Sections

Section 5.1 Analytic Geometry

BASIC CONCEPTS you should learn from this unit.

1. Analytic geometry enables us to combine the ideas of algebra with, or relate the ideas of algebra to, the ideas of geometry.

2. The distance formula is

$$d = \sqrt{(x_2 - x_1)^2 + (y_2 - y_1)^2}$$

3. The midpoint formula

$$x = \frac{x_1 + x_2}{2}; \quad y = \frac{y_1 + y_2}{2}$$

yields the (x, y) coordinates of the midpoint of a line segment joining two points $P_1(x_1, y_1)$ and $P_2(x_2, y_2)$.

WHAT YOU SHOULD KNOW HOW TO DO upon completion of this unit.

1. State the distance formula and the midpoint formula.

2. Use the two formulas to find the distance between two points and the midpoint of a line segment.

3. Use the formulas to assist in the proof of theorems from plane geometry.

SOLUTIONS to Exercise Set 5.1

2. Given (1, 1) and (-2, 5) , the coordinates of the midpoint would be

$$x = \frac{1 + (-2)}{2} = -\frac{1}{2}, \quad \text{and} \quad y = \frac{1 + 5}{2} = 3$$

The midpoint is $(-\frac{1}{2}, 3)$.

5. In similar fashion for the line segment joining (-2, 1) and (-5, -3) we have

$$x = \frac{-2 - 5}{2} = -\frac{7}{2} \quad \text{and} \quad y = \frac{1 - 3}{2} = -1$$

The midpoint is $(-\frac{7}{2}, -1)$.

8. In similar fashion for the points (1, -3), (3, 2) we obtain

$$x = \frac{1 + 3}{2} = 2, \quad y = \frac{-3 + 2}{2} = -\frac{1}{2}$$

The midpoint is $(2, -\frac{1}{2})$.

11. In similar fashion for (1, -1) and (-1, 1) we obtain

$$x = \frac{1 - 1}{2} = 0, \quad y = \frac{-1 + 1}{2} = 0$$

and the midpoint is (0, 0).

14. First, position the rectangle as in this diagram, labeling the midpoints M, P, Q, R, as shown. Since M, P, Q, and R are midpoints of the sides, we can find their coordinates by the midpoint formula.

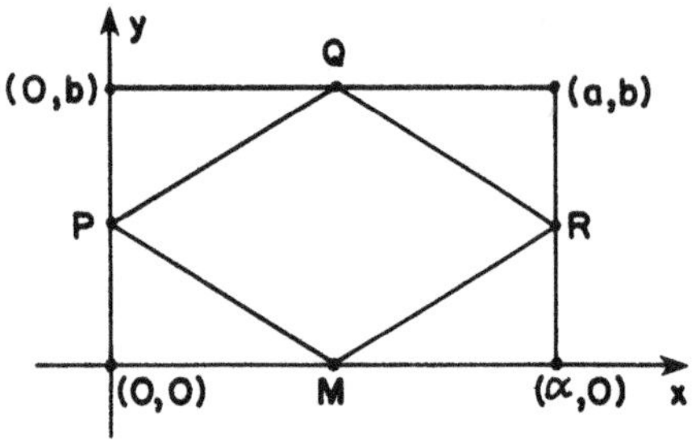

Thus for M

$$x = \frac{0 + a}{2}, \quad y = \frac{0 + 0}{2}; \quad M\left(\frac{a}{2}, 0\right)$$

for P

$$x = \frac{0 + 0}{2}, \quad y = \frac{0 + b}{2}; \quad P\left(0, \frac{b}{2}\right)$$

for Q

$$x = \frac{0 + a}{2}, \quad y = \frac{b + b}{2}; \quad Q\left(\frac{a}{2}, b\right)$$

and for R

$$x = \frac{a + a}{2}, \quad y = \frac{0 + b}{2}; \quad R\left(a, \frac{b}{2}\right)$$

Now

$$\overline{MP} = \sqrt{\left(\frac{a}{2} - 0\right)^2 + \left(0 - \frac{b}{2}\right)^2} = \frac{\sqrt{a^2 + b^2}}{2}$$

$$\overline{PQ} = \sqrt{\left(0 - \frac{a}{2}\right)^2 + \left(\frac{b}{2} - b\right)^2} = \frac{\sqrt{a^2 + b^2}}{2}$$

$$\overline{QR} = \sqrt{\left(\frac{a}{2} - a\right)^2 + \left(b - \frac{b}{2}\right)^2} = \frac{\sqrt{a^2 + b^2}}{2}$$

$$\overline{RM} = \sqrt{\left(a - \frac{a}{2}\right)^2 + \left(\frac{b}{2} - 0\right)^2} = \frac{\sqrt{a^2 + b^2}}{2}$$

Noting that all four line segments are equal, we do have a rhombus formed by MPQR .

17. Place the rectangle as shown in this diagram and show that $\overline{AC} = \overline{BD}$ by the distance formula.

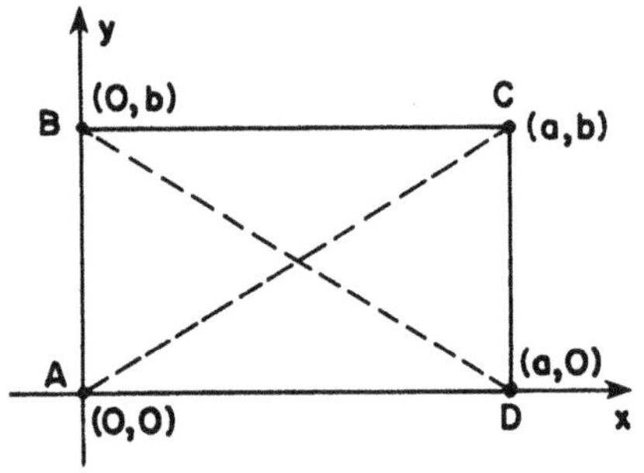

$$\overline{AC} = \sqrt{(0 - a)^2 + (0 - b)^2} = \sqrt{a^2 + b^2}$$

$$\overline{BD} = \sqrt{(0 - a)^2 + (b - 0)^2} = \sqrt{a^2 + b^2}$$

Thus the diagonals are equal.

Section 5.2 The Circle

BASIC CONCEPTS you should learn from this unit.

1. The term conic section comes from the fact that when a plane is passed through a cone at various angles, four different geometric figures are determined: a circle, a parabola, an ellipse, and a hyperbola.

2. Definition of a circle as the set of all points in a plane equidistant from a fixed point. The fixed point is called the center and the given distance the radius.

3. The standard form of the equation of a circle is $(x - h)^2 + (y - k)^2 = r^2$, with center (h, k) and radius r.

WHAT YOU SHOULD KNOW HOW TO DO upon completion of this unit.

1. Define a circle.

2. Write the general equation of a circle in standard form.

3. Write the specific equation of a circle with a given radius and given center.

4. Rewrite into standard form the equation of a circle given in general form (the general form being $Ax^2 + Ay^2 + Dx + Ey + F = 0$, $A \neq 0$), by completing the square method.

5. Determine the center and radius of a circle from its equation in standard form or in general form.

SOLUTIONS to Exercise Set 5.2

2. From the standard form of the equation of the circle we have, for $(h, k) = (-3, 0)$ and $r = 3$,

   $$[x - (-3)]^2 + (y - 0)^2 = 3^2 \quad \text{or} \quad x^2 + 6x + 9 + y^2 = 9$$

   or finally

   $$x^2 + 6x + y^2 = 0$$

5. Similarly, for $(h, k) = (0, 0)$, $r = 3$, we obtain

   $$(x - 0)^2 + (y - 0)^2 = 3^2 \quad \text{or} \quad x^2 + y^2 = 9$$

8. Also, for $(h, k) = (2, 2)$, $r = 2$, we obtain

   $$(x - 2)^2 + (y - 2)^2 = 4$$

11. Given the standard form of the equation of $(x - h)^2 + (y - k)^2 = r^2$ we can read directly from the problem, $(x - 2)^2 + (y + 2)^2 = 4$, that $h = 2, k = -2, r = 2$. Thus the center is $(2, -2)$ and the radius is 2.

14. Given $x^2 + (y - 2)^2 = 4$. To see this one, think of x^2 as $(x - 0)^2$. Then $h = 0$, $k = 2$, and $r = 2$; the center is $(0, 2)$, the radius 2.

17. Rearrange the given equation to read $x^2 + 4x + 4 + y^2 - 8y = 0$. Next, complete the square on the y variable, obtaining this equation:

$$x^2 + 4x + 4 + y^2 - 8y + 16 = 16$$

Then rewrite as

$$(x + 2)^2 + (y - 4)^2 = 4^2$$

The center is $(-2, 4)$, the radius is 4.

20. Dividing the given equation by 2 and rearranging, we obtain

$$x^2 + 4x + y^2 - 6y = 4$$

Now complete the square:

$$x^2 + 4x + 4 + y^2 - 6y + 9 = 4 + 4 + 9$$

and so

$$(x + 2)^2 + (y - 3)^2 = 17$$

The center is $(-2, 3)$; the radius $\sqrt{17}$.

23. Dividing by 3 and rearranging yields $x^2 - 4x + y^2 + 6y = 5$, and completing the square gives

$$x^2 - 4x + 4 + y^2 + 6y + 9 = -5 + 4 + 9$$

Then $(x - 2)^2 + (y + 3)^2 = 8$; the center is $(2, -3)$, the radius $\sqrt{8} = 2\sqrt{2}$.

26. Rearranging the given equation yields $x^2 - 4x + y^2 - 6y = 13$ and completing the square gives us

$$x^2 - 4x + 4 + y^2 - 6y + 9 = 13 + 4 + 9$$

or $(x - 2)^2 + (y - 3)^2 = 26$. This is thus a circle.

31. Divide by 2 and rearrange to get $x^2 - 3x + y^2 - 2y = 1$; completing the square gives

$$x^2 - 3x + \frac{9}{4} + y^2 - 2y + 1 = 1 + \frac{9}{4} + 1$$

and then

$$\left(x - \frac{3}{2}\right)^2 + (y - 1)^2 = \frac{17}{4}$$

which is a circle.

34. Rearrange to read $x^2 + x + y^2 + y = 0$; then

$$x^2 + x + \frac{1}{4} + y^2 + y + \frac{1}{4} = \frac{1}{4} + \frac{1}{4}$$

and thus

$$\left(x + \frac{1}{2}\right)^2 + \left(y + \frac{1}{2}\right)^2 = \frac{1}{2}$$

which is a circle.

Section 5.3 The Parabola

BASIC CONCEPTS you should learn from this unit.

1. Definition of a parabola as the set of all points equidistant from a point and a line, the given point being called the focus and the line being called the directrix.

2. The axis of the parabola is the line through the focus and perpendicular to the directrix.

3. The vertex of the parabola is the point where the axis intersects the parabola.

4. If the distance from the vertex to the focus of a parabola is p units and the vertex is at the origin, then the equation of the parabola is
   A. $x^2 = 4py$ and the graph has y-axis symmetry, or
   B. $y^2 = 4px$ and the graph has x-axis symmetry.

   For simplicity we let $a = 4p$; then the equations become $x^2 = ax$ and $y^2 = ay$.

5. If $a > 0$ in the equations in concept 4, the graph opens upward for $x^2 = ay$ and to the right for $y^2 = ax$. When $a < 0$, the directions of opening are downward and to the left, respectively.

6. The standard forms of the equations of the parabola generally are
   A. $(x - h)^2 = a(y - k)$
   in which case the vertex is (h, k), axis $x = h$, and the direction is upward if $a > 0$, downward if $a < 0$; and
   B. $(y - k)^2 = a(x - h)$
   in which case the vertex is (h, k), axis $y = k$, and the direction is to the right if $a > 0$, to the left if $a < 0$.

7. The method of completing the square can be used to write the equation of a parabola in its standard form.

WHAT YOU SHOULD KNOW HOW TO DO upon completion of this unit.

1. Define the terms parabola, vertex, focus, axis, and directrix.

2. Rewrite the equation of a parabola into the standard form.

3. From the standard form of the parabola equation, determine the vertex, axis, direction, focus, and directrix.

4. Graph a parabola from its equation.

SOLUTIONS to Exercise Set 5.3

2. $x^2 = -4y$

a < 0; the direction is downward;

the axis is the y-axis.

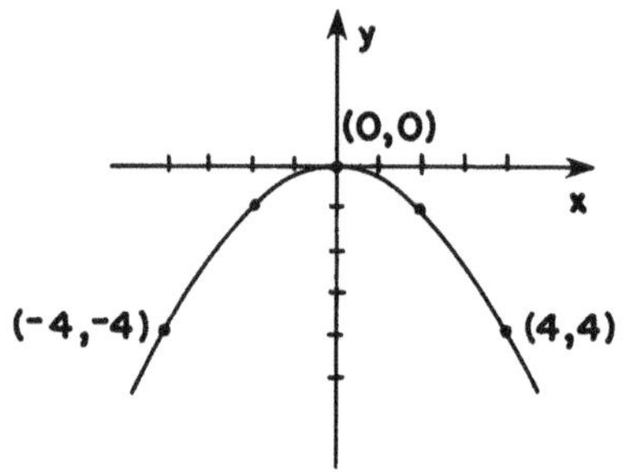

5. $x^2 + 5y = 0$

$x^2 = -5y$

A < 0; the direction is downward;

the axis is the y-axis.

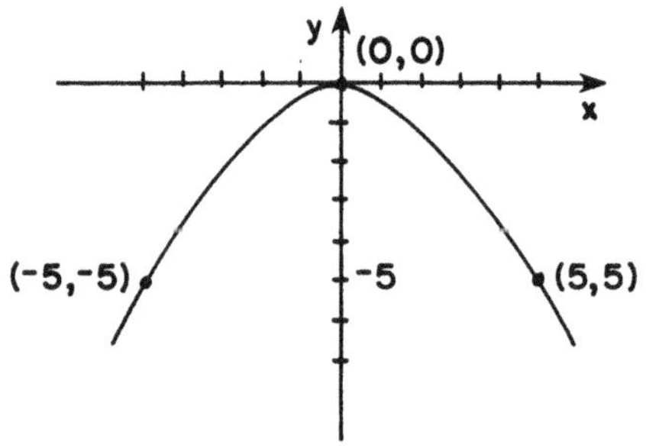

8. $(y - 2)^2 = -2(x + 1)$

The vertex is at (-1, 2);

a < 0; the direction is to the left;

the axis y = 2 .

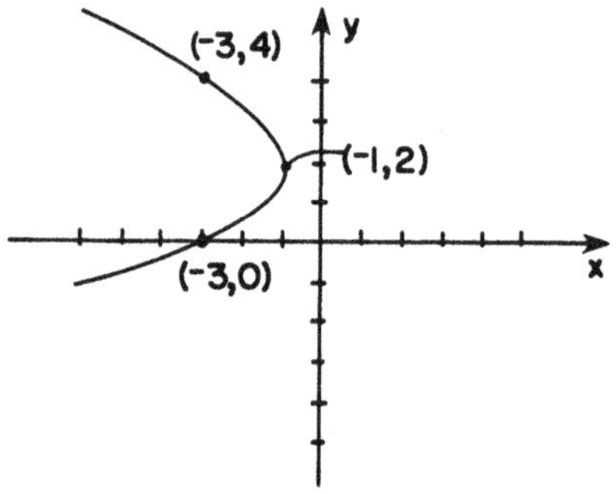

11. $y^2 = -2(x + 1)$

The vertex is at (-1, 0);

the axis is the x-axis;

a < 0, so the direction is to the left.

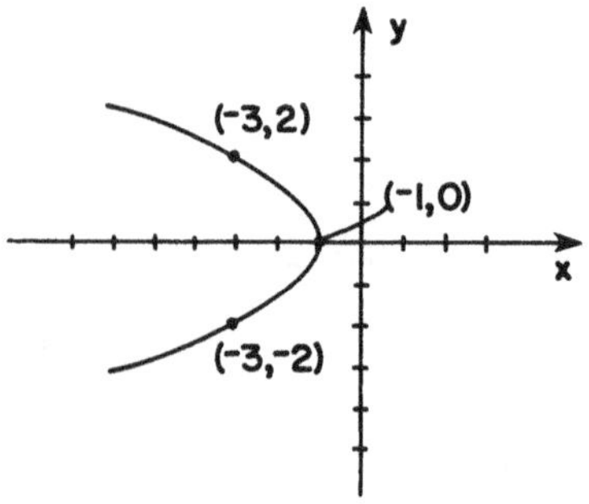

14. $x^2 + 4x + 2y - 2 = 0$. Rearranging and completing the square, we have $x^2 + 4x + 4 = -2y + 2 + 4$; $(x + 2)^2 = -2(y - 3)$. From this we determine the vertex to be $(-2, 3)$, the axis to be $x = -2$, and the direction to be downward.

17. $x^2 - x + 3y + 1 = 0$. Rearranging and completing the square yields $x^2 - x + 1/4 = -3y - 1 + 1/4$; $(x - 1/2)^2 = -3(y + 1/4)$. The vertex is at $(1/2, -1/4)$, the axis $x = 1/2$, and the direction is down.

20. $x^2 + 2x - 5y - 19 = 0$. Rearranging and completing the square yields $x^2 + 2x + 1 = 5y + 20$ or $(x + 1)^2 = 5(y + 4)$. The vertex is at $(-1, -4)$, the axis $x = -1$, and the direction is up.

23. $y^2 + 6y + \frac{1}{2}x + 7 = 0$. Rearranging and completing the square yields $y^2 + 6y + 9 = -\frac{1}{2}x - 7 + 9$ or $(y + 3)^2 = -\frac{1}{2}(x - 4)$. The vertex is at $(4, -3)$, the axis $y = -3$, and the direction is to the left.

26. $y^2 - 6y + 2x + 17 = 0$. Rearranging and completing the square yields $y^2 - 6y + 9 = -2x - 17 + 9$ or $(y - 3)^2 = -2(x + 4)$. The vertex is at $(-4, 3)$, the axis $y = 3$; the graph opens to the left.

Section 5.4 The Ellipse and Hyperbola

BASIC CONCEPTS you should learn from this unit.

1. Definition of an ellipse as the set of all points the sum of whose distances from two fixed points is a constant, the two points being called the foci.

2. The standard form of the equation of an ellipse that is centered at the origin is

$$\frac{x^2}{a^2} + \frac{y^2}{b^2} = 1$$

with $\pm a$ being the x-intercepts and $\pm b$ being the y-intercepts.

3. Definition of a hyperbola as the set of all points the difference of whose distances from two fixed points is a constant, the two points being called the foci.

4. The standard forms of the equation of a hyperbola that is centered at the origin are

A. $\dfrac{x^2}{a^2} - \dfrac{y^2}{b^2} = 1$

(In this case the foci are on the x-axis and x-intercepts are $\pm a$);

B. $\dfrac{y^2}{a^2} - \dfrac{x^2}{b^2} = 1$

(In this case the foci are on the y-axis and y-intercepts are $\pm a$).

5. The asymptotes of a hyperbola of the form in concept 4A are lines with equations

$$y = \pm\frac{b}{a}x$$

The asymptotes of a hyperbola of the form in concept 4B are lines with equations

$$y = \pm\frac{a}{b}x$$

WHAT YOU SHOULD KNOW HOW TO DO upon completion of this unit.

1. Recognize the equation of an ellipse in its standard form.
2. Graph an ellipse from its equation.
3. Rewrite the equation of an ellipse into the standard form.
4. Recognize the equation of a hyperbola in the standard form.
5. Graph a hyperbola from its equation, using intercepts.
6. Write the equations of the asymptotes of a hyperbola and use them in graphing the curve.
7. Rewrite the equation of a hyperbola into the standard form.

SOLUTIONS to Exercise Set 5.4

2. $\frac{x^2}{4} + \frac{y^2}{16} = 1$ (2, 0), (-2, 0) are the x-intercepts and (0, 4), (0, -4) the y-intercepts.

The graph is

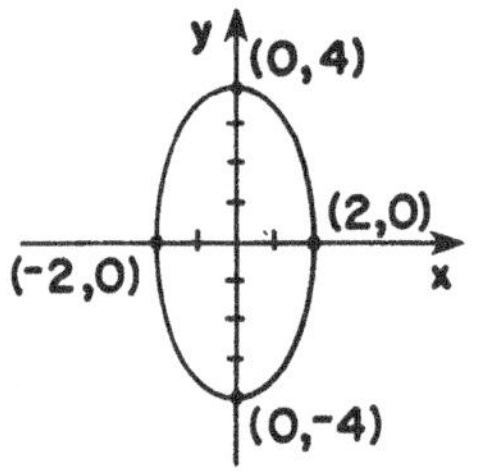

5. $\frac{x^2}{16} + \frac{y^2}{25} = 1.$ (4, 0), (-4, 0) are the x-intercepts and (0, 5), (0, -5) the y-intercepts;

the graph is

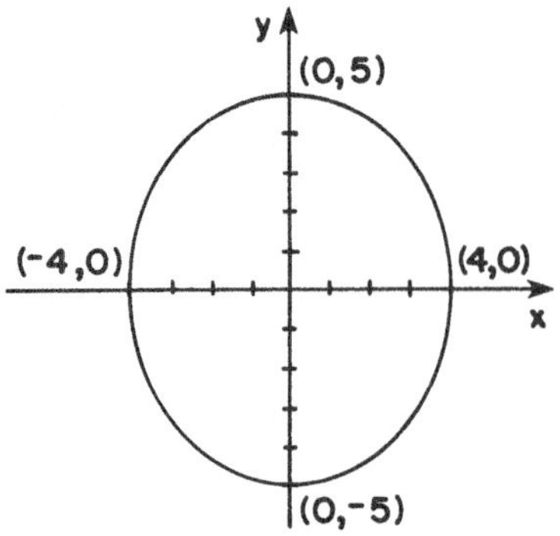

8. $16x^2 + 9y^2 = 144$. Dividing by 144 yields the standard form

$$\frac{x^2}{9} + \frac{y^2}{16} = 1$$

and from this we see the x-intercepts, (3, 0), (-3, 0), and the y-intercepts, (0, 4), (0, -4).

11. $4x^2 + 16y^2 = 4$. Dividing by 4 gives us the standard form

$$\frac{x^2}{1} + \frac{y^2}{\frac{1}{4}} = 1$$

The x-intercepts are (1, 0), (0, -1); the y-intercepts are $\left(0, \frac{1}{2}\right)$, $\left(0, -\frac{1}{2}\right)$.

14. $5x^2 + 6y^2 = 50$. Dividing by 50 and rewriting $\frac{6}{50}$ as $\frac{25}{3}$ in the denominator yields the standard form,

$$\frac{x^2}{10} + \frac{y^2}{\frac{25}{3}} = 1$$

and thus the x-intercepts are $(\sqrt{10}, 0)$, $(-\sqrt{10}, 0)$ and the y-intercepts are $\left(0, \frac{5\sqrt{3}}{3}\right)$, $\left(0, \frac{-5\sqrt{3}}{3}\right)$.

17. $\frac{x^2}{25} - \frac{y^2}{16} = -1$ rewritten in standard form is

$$\frac{y^2}{16} - \frac{x^2}{25} = 1$$

and thus the intercepts are (0, 4) and (0, -4). The graph is thus

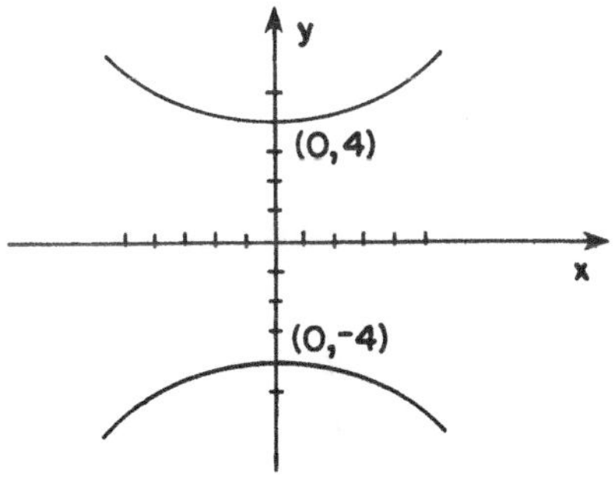

20. $$\frac{y^2}{49} - \frac{x^2}{25} = 1$$

is already in standard form and the intercepts are (0, 7), (0, -7). The graph is

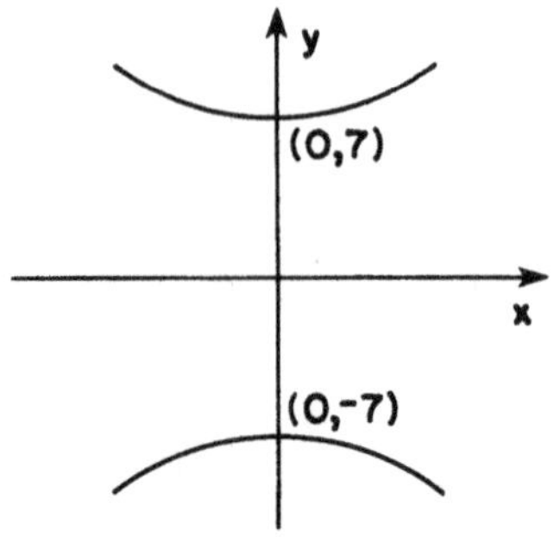

23. $16x^2 - y^2 = 64$. Dividing by 64 yields

$$\frac{x^2}{4} - \frac{y^2}{64} = 1$$

The intercepts are (2, 0), (-2, 0).

26. $2x^2 - 3y^2 = 6$. Dividing by 6 yields

$$\frac{x^2}{3} - \frac{y^2}{2} = 1$$

The intercepts are $(\sqrt{3}, 0)$, $(-\sqrt{3}, 0)$.

29. $16x^2 - 9y^2 = 144$. First divide by 144 to put the equation in standard form:

$$\frac{x^2}{9} - \frac{y^2}{16} = 1$$

with the $x^2/9$ term positive, the asymptotes are

$$y = \pm\frac{b}{a}x \quad \text{or} \quad y = \frac{4}{3}x \quad \text{and} \quad y = -\frac{4}{3}x$$

and the graph is

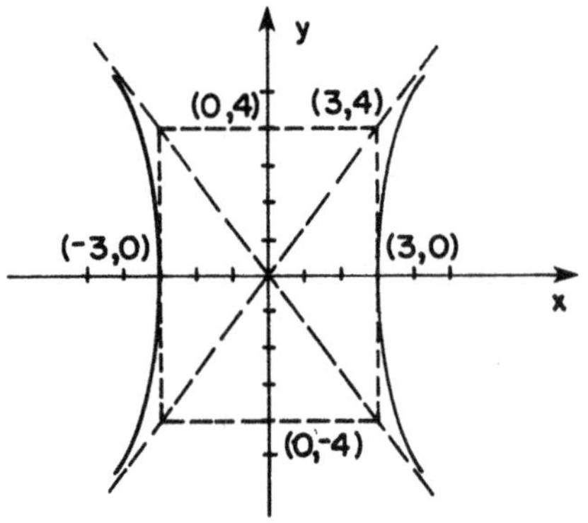

32. $25x^2 - 9y^2 = 225$ Divide by 225

$$\frac{x^2}{9} - \frac{y^2}{25} = 1$$

The asymptotes will be

$$y = \pm\frac{b}{a}x \qquad \text{or} \qquad y = \pm\frac{5}{3}x$$

The graph is

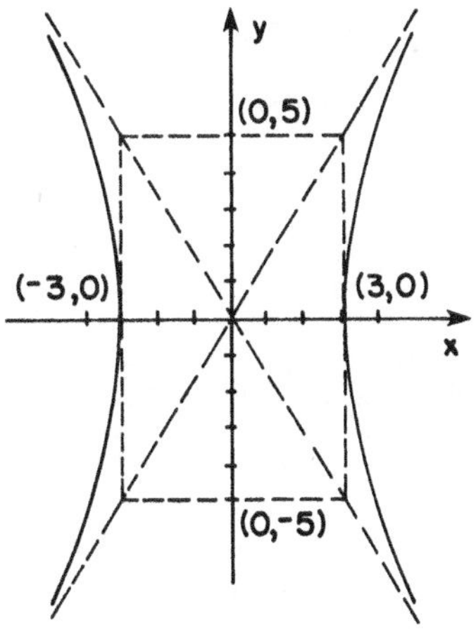

35. Referring to the figure

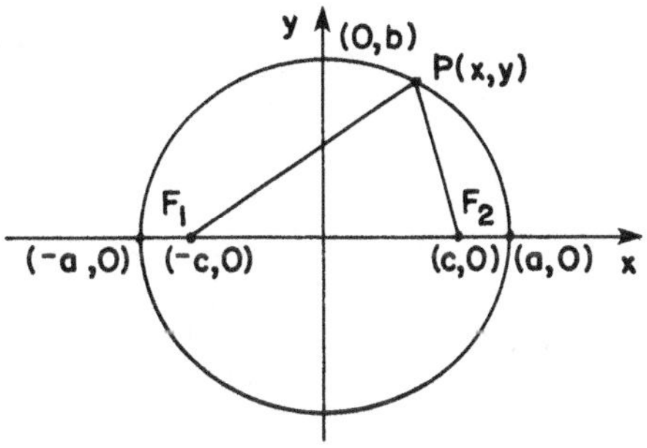

we see that $F_1P + PF_2 = 2a$. By the distance formula this relation can be written as

$$\sqrt{(x + c)^2 + y^2} + \sqrt{(x - c)^2 + y^2} = 2a$$

or

$$\sqrt{(x + c)^2 + y^2} = 2a - \sqrt{(x - c)^2 + y^2}$$

Then, squaring, we obtain

$$x^2 + 2xc + c^2 + y^2 = 4a^2 - 4a\sqrt{(x - c)^2 + y^2} + x^2 - 2xc + c^2 + y^2$$

or $4xc - 4a^2 = -4a\sqrt{(x - c)^2 + y^2}$. Dividing by 4 gives $xc - a^2 = -a\sqrt{(x - c)^2 + y^2}$,

squaring again, we get

$$x^2c^2 - 2a^2xc + a^4 = a^2x^2 - 2a^2xc + a^2c^2 + a^2y^2$$

35. (continued)

or $x^2(a^2 - c^2) + a^2y^2 = a^2(a^2 - c^2)$. Dividing by $a^2(a^2 - c^2)$ yields

$$\frac{x^2}{a^2} + \frac{y^2}{a^2 - c^2} = 1$$

Now substitute $b^2 = a^2 - c^2$ to get the standard form:

$$\frac{x^2}{a^2} - \frac{y^2}{b^2} = 1$$

Section 5.5 Identifying the Conic Sections

BASIC CONCEPTS you should learn from this unit.

1. There are two basic forms of the equation of a conic section:

   A. The general equation

   $$Ax^2 + Cy^2 + Dx + Ey + F = 0$$

   B. The standard form equation, which is obtained from the foregoing by completing the square.

2. The various conic sections can be identified from the general equation as follows.

   A. Either A or $C = 0$ (not both): Parabola.

   B. $A = C$: Circle.

   C. $A \neq C$ and $AC > 0$: Ellipse.

   D. $A \neq C$ and $AC < 0$: Hyperbola.

3. When the equation is in standard form much valuable information about the conic sections is seen by inspection: the location of the center or vertex; the location of the intercepts; the axis; the direction of opening; and the asymptotes. (See Table 2 in the text for details.)

WHAT YOU SHOULD KNOW HOW TO DO upon completion of this unit.

1. Identify a conic section as a circle, parabola, ellipse, or hyperbola by analyzing the <u>general</u> equation.

SOLUTIONS to Exercise Set 5.5

2. $-x^2 + 4y^2 + 2x - 3y + 5 = 0$. Here $A \neq C$ and $AC = -5 < 0$, so the conic section is a <u>hyperbola</u>.

5. $36x^2 - y^2 + x - y + 2 = 0$. Here, too, $A \neq C$ and $AC < 0$, so the conic section is a <u>hyperbola</u>.

8. $2y^2 - 3x + y + 4 = 0$. $A = 0$, so the conic section is a <u>parabola</u>.

11. $4x^2 + 9y^2 - x + 2 = 0$. $A \neq C$ and $AC > 0$. The conic section might be an ellipse, but completing the square reveals that

$$\left(x^2 - \frac{1}{4}x + \frac{1}{64}\right) + \frac{9}{4}y^2 = -\frac{1}{2} + \frac{1}{64} = -\frac{31}{64}$$

and since $-\frac{31}{64}$ is nonpositive, we have no graph.

14. $x^2 + y^2 + 6x - 2y + 10 = 0$. $A = C$; we might have a circle. Check by completing the square:

$$(x^2 + 6x + 9) + (y^2 - 2y + 1) = -10 + 9 + 1 = 0$$

So we have a circle with zero radius or a point. The point is the center: $(-3, 1)$.

OLUTIONS to Review Exercises Chapter 5

2. Using the midpoint formula,

$$x_m = \frac{-2 - 3}{2} = -\frac{5}{2} \quad \text{and} \quad y_m = \frac{0 + 5}{2} = \frac{5}{2}$$

The midpoint is $\left(-\frac{5}{2}, \frac{5}{2}\right)$.

5. $P_1P_2 = \sqrt{(3 - 2)^2 + (4 + 1)^2} = \sqrt{26}$, $\quad P_2P_3 = \sqrt{(-1 - 3)^2 + (1 - 4)^2} = 5$

$P_3P_4 = \sqrt{(-2 + 1)^2 + (-4 - 1)^2} = \sqrt{26}$, $\quad P_4P_1 = \sqrt{(-2 - 2)^2 + (-4 + 1)^2} = 5$

Since opposite sides are equal, the quadrilateral is a parallelogram.

8. $y^2 = x^2 - 9$. Substituting $-x$ for x and $-y$ for y leaves the equation unchanged, so we have symmetry with both axes and the origin.

11. $3x^2 - 2x - 4y^2 = 6$. Replacing x by $-x$ changes the $-2x$ term to $+2x$, so there is no symmetry with the y-axis. But replacing $-y$ for y does not change the equation, so there is symmetry with the x-axis.

14. The standard form of the circle equation is $(x - h)^2 + (y - k)^2 = r^2$. Substituting the given values yields $(x + 3)^2 + (y + 3)^2 = 4$.

17. $x^2 + y^2 + 4x - 6y = -10$. Completing the square gives

$$x^2 + 4x + 4 + y^2 - 6y + 9 = -10 + 4 + 9$$

or

$$(x + 2)^2 + (y - 3)^2 = 3$$

from which we read the center to be $(-2, 3)$ and the radius $= \sqrt{3}$.

20. $x^2 + y^2 - 2x - 2y = -8$. Completing the square gives

$$x^2 - 2x + 1 + y^2 - 2y + 1 = -8 + 1 + 1$$

or

$$(x - 1)^2 + (y - 1)^2 = -6$$

which is impossible.

23. $y^2 + 3x + 9 = 0$ in standard form is $y^2 = 3(x + 3)$. The vertex thus is $(-3, 0)$, the axis is $y = 0$, and since $a = +3$, the parabola opens to the right.

26. $x^2 + 4x + 2y + 5 = 0$. Completing the square and rearranging gives $x^2 + 4x + 4 = -2y - 5 + 4$ or $(x + 2)^2 = -2(y + 1/2)$. Thus the vertex is $(-2, -1/2)$, the axis is $x = -2$, and $a = -2$, which is less than 0, so the direction of the opening is downward.

29. $9x^2 - 4y^2 = 36$ in standard form is

$$\frac{x^2}{4} - \frac{y^2}{9} = 1$$

This hyperbola intercepts the x-axis at $(\pm 2, 0)$.

32. $9x^2 - 16y^2 = 144$ in standard form is

$$\frac{x^2}{16} - \frac{y^2}{9} = 1$$

This hyperbola intercepts the x-axis at $(\pm 4, 0)$.

35. $4x^2 - 4y^2 = 1$. Rearrange to

$$\frac{x^2}{1/4} - \frac{y^2}{1/4} = 1$$

to see that $a^2 = (1/2)^2$ and $b^2 = (1/2)^2$. Intercepts are found by setting $y = 0$; then $x = \pm 1/2$, and the intercepts are at $(\pm 1/2, 0)$. The asymptotes are $y = \pm b/a\, x$ or $y = \pm x$.

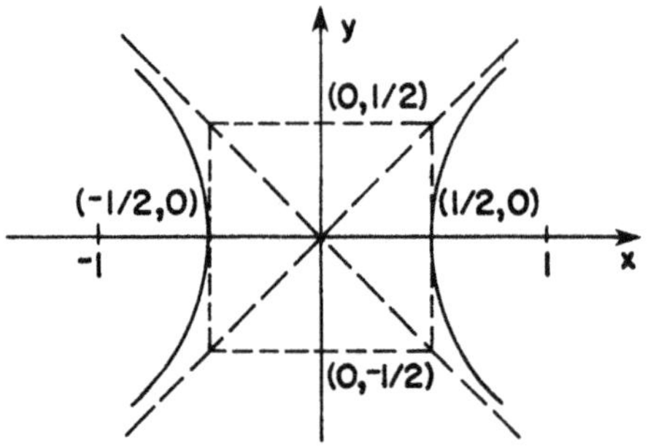

38. $6x^2 - 7y^2 - 5x + 6y = 0$. Using the criteria in Section 8.5, we see that $A = 6$ and $C = -7$, so $A \neq C$ and $AC < 0$; hence the conic section is a hyperbola.

Chapter 5 - A SHORT QUIZ

1. State the midpoint formula.

2. $(x - h)^2 + (y - k)^2 = r^2$ is the standard form of which conic section?

3. In Question 2, what are h, k, and r ?

4. Which conic section is represented by

$$\frac{x^2}{a^2} - \frac{y^2}{b^2} = 1 \quad ?$$

5. In Question 4, which axis is intersected?

6. $(y - 2)^2 = -(x - 3)$ is the equation of a ___?___ whose axis is parallel to which axis? (x-axis or y-axis)

7. Write the equation of a hyperbola in standard form that intercepts the y-axis at $(0, \pm 2)$.

8. Write an equation of a parabola that is symmetric to the y-axis and opens downward with its vertex at (0, 1) .

9. Which axis is intersected by $x^2 = 4(y - 2)$?

10. What is the direction of opening of the graph of the equation in Question 9?

11. What are the x- and y-intercepts of

$$\frac{x^2}{2} + \frac{y^2}{9} = 1 \quad ?$$

12. What type of symmetry is present in the graph of Question 11?

13. What is the relationship between a and b in the standard form of the ellipse equation if the ellipse is centered at the origin and has its longer axis along the y-axis?

14. True or False: The focus of a parabola lies on the directrix.

15. True or False: All points on a parabola are equidistant from the focus and the directrix.

16. True or False: The axis of a parabola is a line through the vertex and parallel to the directrix.

17. Write the equation of the directrix for this parabola: $y^2 = -4(x - 1)$.

18. In the standard form of the equation of a parabola, $(x - h)^2 = a(y - k)$, what do h and k represent?

19. Where does a hyperbola intersect its asymptotes?

20. Write the equation of a conic section such that the graph will be a single point.

Quiz Answers Chapter 5

1. $x_m = \dfrac{x_1 + x_2}{2}$; $y_m = \dfrac{y_1 + y_2}{2}$

2. Circle

3. (h, k) are the coordinates of the center, r is the radius.

4. Hyperbola

5. x-axis

6. Parabola; x-axis

7. $\frac{y^2}{4} - \frac{x}{7} = 1$

8. $x^2 = (y - 1)$

9. y-axis

10. upward

11. The x-intercepts are $(\pm\sqrt{2}, 0)$; the y-intercepts are $(0, \pm 3)$.

12. Both axes and origin

13. $b > a$

14. False

15. True

16. False

17. $x = 2$

18. Coordinates of the vertex

19. It does <u>not</u> intersect the asymptotes.

20. $(x - 2)^2 + y^2 = 0$ (Any circle with radius = 0 will do.)

CHAPTER 6

Systems of Equations and Inequalities

Section 6.1 Systems of Linear Equations

BASIC CONCEPTS you should learn from this unit.

1. A system of linear equations is a pair of linear equations with two unknowns.
2. Solving a system of equations means finding values of the unknowns that satisfy both equations.
3. A system that has a solution is said to be consistent. A system with no solution is inconsistent.
4. A system can be solved by graphing, in which case the solution is the coordinates of the point where the two lines intersect.
5. There are three possible results when two linear equations are graphed:
   A. Two different lines intersect once;
   B. Two parallel lines are obtained;
   C. Both equations graph onto the same line.
6. The substitution method of solving a system yields an exact solution, whereas graphing is approximate.
7. If the substitution method results in a mathematical inconsistency, the system has no solution.
8. If the method results in a mathematical identity, the system has many solutions.

WHAT YOU SHOULD KNOW HOW TO DO upon completion of this unit.

1. Define and recognize a system of linear equations.
2. Solve a system of two linear equations by graphing, and algebraically by substitution.
3. Define the terms "consistent system" and "inconsistent system."
4. Illustrate graphically a consistent system and an inconsistent system.

SOLUTIONS to Exercise Set 6.1

2. (1) $x - y = 1$
   (2) $x + y = 5$

(1)

x	y
1	0
3	2

(2)

x	y
1	4
4	1

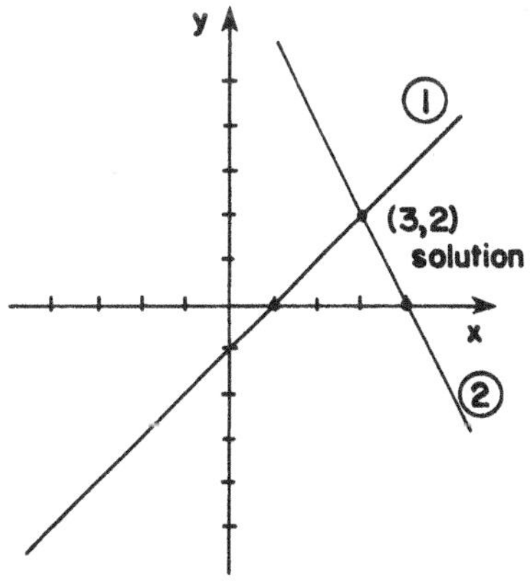

5. (1) $x + 3y = -2$

(2) $3x - 5y = 8$

(1)

x	y
-2	0
1	-1

(2)

x	y
1	-1
-4	-4

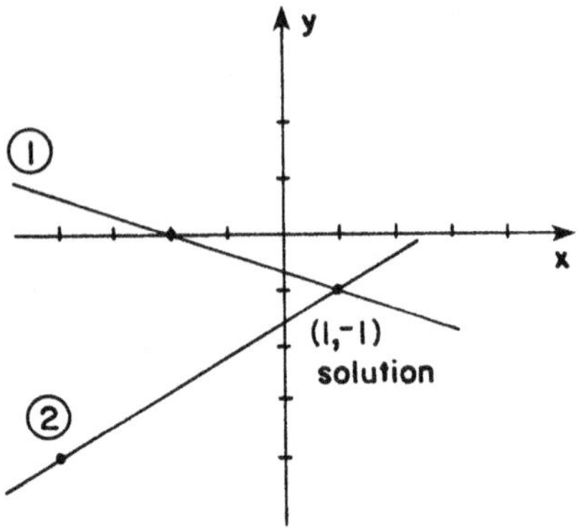

8. (1) $3x + 2y = 10$

(2) $-9x - 6y = 8$

(1)

x	y
0	5
4	-1

(2)

x	y
0	-4/3
-2	5/3

Lines appear parallel.

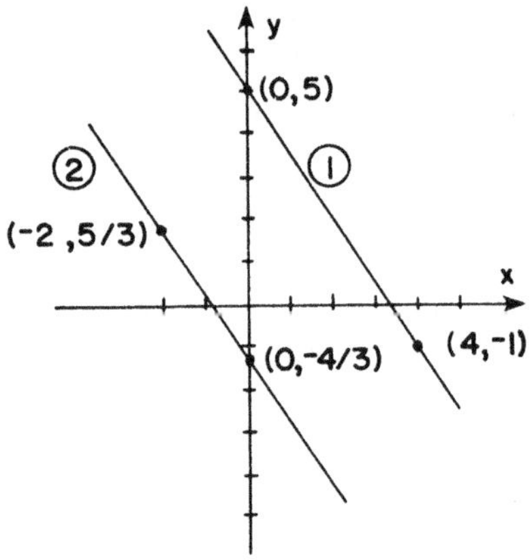

11. (1) $x + 3y = 3$

(2) $x - 5y = -1$

From (1) $x = -3y + 3$

Now substitute in (2), obtaining

$$(-3y + 3) - 5y = -1$$
$$-8y = -4$$
$$y = 1/2$$

$$x = -3y + 3$$
$$x = -3\left(\frac{1}{2}\right) + 3$$
$$x = 3/2$$

14. (1) $3x - y = -9$

(2) $2x - y = -7$

From (1) $y = 3x + 9$

Now substituting in (2) we get

$$2x - (3x + 9) = -7$$
$$-x - 9 = -7$$
$$x = -2$$

$$y = 3(-2) + 9$$
$$y = 3$$

17. (1) $3x - y = 18$ From (1) $y = 3x - 18$

(2) $\frac{3}{2}x - \frac{1}{2}y = 9$ and substituting in (2) yields

$$\frac{3}{2}x - \frac{1}{2}(3x - 18) = 9$$

$$3x - (3x - 18) = 18$$

$0 = 0$ an identity, thus any solution of (1) is a solution of (2).

Section 6.2 Solving by Elimination

BASIC CONCEPTS you should learn from this unit.

1. The method of solving a system of two linear equations by elimination consists of manipulating the coefficients of the variables so that when the two equations are added one variable vanishes, and the resulting equation has only one variable and is easily solved.

2. If the method results in $0x + 0y = C$, $c \neq 0$, there is an inconsistent system. (no solution).

3. If the method results in $0x + 0y = 0$, then the two equations were essentially the same and there are many solutions.

WHAT YOU SHOULD KNOW HOW TO DO upon completion of this unit.

1. Solve a pair of linear equations by the method of elmination.

2. Detect an inconsistent system by the result of this method.

3. Determine that the two equations were essentially the same by the result of this method.

SOLUTIONS to Exercise Set 6.2

2.
$$x - 2y = 8$$
$$2x + y = 1$$

Multiply the second equation by 2, obtaining a new system:

$$x - 2y = 8$$
$$4x + 2y = 2$$

Adding gives $5x = 10$ and $x = 2$. Now substituting 2 for x in the first equation gives $2 - 2y = 8$ and $y = -3$.

5.

$$\begin{array}{r} x + 2y = 6 \\ 2x + 4y = 12 \end{array}$$

Multiply the first equation by -2 to obtain a new system, then add

$$\begin{array}{r} -2x - 4y = -12 \\ 2x + 4y = 12 \\ \hline 0x + 0x = 0 \end{array}$$

which is always true. Thus there are many solutions.

8.

$$\begin{array}{r} x + 2y = 0 \\ 5x - y = 22 \end{array}$$

Multiply the second equation by 2, obtaining a new system. Then add.

$$\begin{array}{r} x + 2y = 0 \\ 10x - 2y = 44 \\ \hline 11x \quad = 44 \end{array}$$

Hence $x = 4$ and substituting 4 for x in the first equation gives $y = -2$.

11.

$$\begin{array}{r} x + 2y = 1 \\ 5x + 2y = 13 \end{array}$$

Multiply the first equation by -1, obtaining a new system. Then add.

$$\begin{array}{r} -x - 2y = -1 \\ 5x + 2y = 13 \\ \hline 4x \quad = 12; \end{array} \quad x = 3.$$

Substituting gives $y = -1$.

14.

$$\begin{array}{r} 2x - 3y = 8 \\ 4x - 6y = 16 \end{array}$$

Multiply the first equation by -2, obtaining a new system. Then add.

$$\begin{array}{r} -4x + 6y = -16 \\ 4x - 6y = 16 \\ \hline 0x + 0y = 0 \end{array}$$

which is an identity. So there are many solutions.

17. Let x = number of pounds of nuts
y = number of pounds of raisins

$$x + y = 2 \qquad \text{(We want 2 lb of mix.)}$$

$$2.10x + .90y = 2(1.62) \qquad \text{value of the nuts}$$

plus the value of the raisins equals the value of the mix. Multiply the first equation by -90 and the second equation by 100; we obtain

$$\begin{aligned} -90x - 90y &= -180 \\ \underline{210x + 90y} &= \underline{324} \\ 120x \qquad &= 144 \end{aligned} \qquad \text{Add these equations.}$$

$$x = 144/120 = 6/5 \text{ lb}$$

and substituting gives $6/5 + y = 2$, $y = 4/5$ lb. 6/5 lb of nuts and 4/5 lb of raisins.

20. Let x = the amount invested at 12%, and
y = the amount invested at 10%

$$x + y = 6000 \qquad \text{Sum of the two investments = \$6000.}$$

$$0.12x + 0.10y = 620 \qquad \text{Sum of the interests = \$620}$$

Multiply the first equation by -10 and the second by 100 to obtain

$$\begin{aligned} -10x - 10y &= -60000 \\ \underline{12x + 10y} &= \underline{62000} \\ 2x \qquad &= 2000 \\ x &= 1000 \end{aligned}$$

and clearly $y = \$5000$.

Section 6.3 Applications

BASIC CONCEPTS YOU SHOULD LEARN from this unit.

1. Word problems can often be set up with two equations with two unknowns and then solved.
2. The BREAK-EVEN point in a business is when the number of items sold produces income or revenue exactly equal to their cost.
3. Revenue and costs are often depicted by algebraic equations.
4. In an economic system the equilibrium point is the solution of the demand equation and the supply equation. The ability to solve a system of equations, then, assists us in determining the equilibrium point.

WHAT YOU SHOULD KNOW HOW TO DO upon completion of this unit.

1. Set up and solve word problems using two unknowns and forming two equations.
2. Determine the break-even point in a manufacturing and selling type problem.
3. Determine the equilibrium point in a supply and demand type problem.

SOLUTIONS to Exercise Set 6.3

2\. Let x = the number of ten-cent stamps
y = the number of twenty-cent stamps

Then:

$$x + y = 27$$
$$10x + 20y = 300$$

Multiply the first equation by -10 to obtain

$$\begin{aligned} -10x - 10y &= -270 \\ 10x + 20y &= 300 \\ \hline 10y &= 30 \\ y &= 3 \end{aligned}$$

Add these equations.

and so $x = 24$.

Check: $3 + 24 = 27$

and $10(24) + 20(3) = 300$

5\. Let A = the amount invested in Type A
B = the amount invested in Type B

Then $A + B = 6000$ Sum of the investments = \$6000.

$.08A + .10B = 520$ Sum of the dividends = \$ 520.

$-8A - 8B = -48000$ Multiplying by -8

$8A + 10B = 52000$ Multiplying by 100

$$\begin{aligned} 2B &= 4000 \\ B &= \$\ 2000 \end{aligned}$$

Then $A = \$4000$.

8\. Let a = the number of units of grain A
b = the number of units of grain B

$2a + 3b = 18$ fat equation

$80a + 60b = 480$ calories equation

$$\begin{aligned} 2a + 3b &= 18 \\ -4a - 3b &= -24 \\ \hline -2a \quad &= -6 \end{aligned}$$

$$\begin{aligned} 6 + 3b &= 18 \\ 3b &= 12 \\ b &= 4 \end{aligned}$$

$a = 3$ units of grain A

$b = 4$ units of grain B

11. Let x = the speed of the bike in still air
y = the speed of the wind

Using the formula $d = rt$ we get

$$4(x - y) = 45 \quad \text{and}$$
$$3(x + y) = 45$$

or
$$4x - 4y = 45 \qquad \text{or} \qquad 12x - 12y = 135$$
$$3x + 3y = 45 \qquad\qquad \underline{12x + 12y = 180}$$
$$24x \qquad = 315$$

and $x = \frac{315}{24} = \frac{105}{8}$ mph

and since

$3x + 3y = 45$ or $x + y = 15$, $y = 15 - \frac{105}{8} = \frac{15}{8}$

14. Let t = the tens digit
u = the units digit

Then
$$3t + u = 14$$
$$t + 2u = 18$$

will be the equations, or

$$-6t - 2u = -28$$
$$\underline{t + 2u = 18}$$
$$-5t \qquad = -10$$

and $t = 2$, and from the first equation $6 + u = 14$ and $u = 8$. Thus the number is 28.

17. Let a = the amount invested in model A
b = the amount invested in model B

Then
$$a + b = 18000$$
$$.12a + .18b = (.16)(18000)$$

are the equations; or

$$a + b = 18000 \qquad \text{or} \qquad -2a - 2b = -36000$$
$$2a + 3b = 48000 \qquad\qquad \underline{2a + 3b = 48000}$$
$$b = \$12000$$

Then the first equation yields $a = \$\ 6000$

20. Graphing each equation yields

(a) Break-even point occurs at 8000 kg.

(b) Revenue for 8000 kg:
R = .75(8000) = \$6000

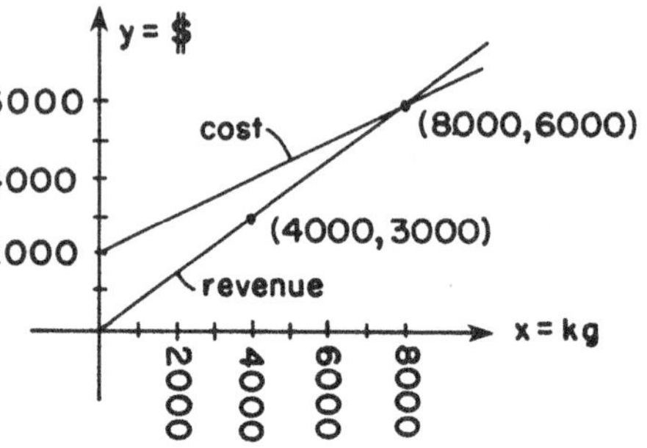

23. a. Equilibrium occurs when S = D or in this case 2p + 10 = -p + 22 . Solving this for p yields p = 4 .

b. The number sold at this price can be found from either equation: s = 2p + 10 = 18 .

Section 6.4   Systems of Linear Equations in Three Unknowns

BASIC CONCEPTS you should learn from this unit.

1. Three linear equations in three unknowns can be solved by the substitution method and by the elimination method.

2. Gaussian elimination is another method of solving systems of linear equations. The method consists of applying these three operations.

   A. Interchange any two equations.

   B. Multiply an equation by a nonzero constant.

   C. Replace an equation by itself plus a constant times another equation.

3. If Gaussian elimination results in

   A. $0x + 0y + 0z = C,\ C \neq 0$, the system is inconsistent;

   B. $0x + 0y + 0z = 0$, the system is consistent and has many solutions.

WHAT YOU SHOULD KNOW HOW TO DO upon completion of this unit.

1. Find the solutions (if any) to a system of three linear equations by Gaussian elimination.

2. If a system has infinitely many solutions, state the general solution.

3. Recognize when a system has no solution.

SOLUTIONS to Exercise Set 6.4

2. $2x + 3y + 4z = -12$

$x - 2y + z = -5$

$3x + y + 2z = 1$ Interchange the first and second equations, obtaining

$x - 2y + z = -5$

$2x + 3y + 4z = -12$ Add -2 times the first equation to the second equation, and

$3x + y + 2z = 1$ add -3 times the first equation to the third equation, obtaining

$x - 2y + z = -5$

$7y + 2z = -2$ Add -1 times the second equation to the third equation,

$7y - z = 16$ obtaining,

$x - 2y + z = -5$

$7y + 2z = -2$

$-3z = 18$

Now from the third equation $z = -6$, and back substitution yields $x = \frac{27}{7}$, $y = \frac{10}{7}$

5. $x + y + z = 2$

$x - y + 2z = 3$ Add -1 times the first equation to the second equation, and

$3x + 5y + 2z = 6$ -3 times the first equation to the third equation, obtaining

$x + y + z = 2$

$-2y + z = 1$

$2y - z = 0$ Add the second equation to the third equation, obtaining

$x + y + z = 2$

$-2y + z = 1$

$0y + 0z = 1$ an inconsistency: so there is no solution.

8. $4x + 2y - z = 5$

$3x + 3y + 6z = 1$ Add -1 times the second equation to the first equation,

$5x + y - 8z = 8$ obtaining

$x - y - 7z = 4$

$3x + 3y + 6z = 1$ Add -3 times the first equation to the second equation, and

$5x + y - 8z = 8$ -5 times the first equation to the third equation, obtaining

$x - y - 7z = 4$

$6y + 27z = -11$

$6y + 27z = -12$ Add -1 times the second equation to the third equation, obtaining

$x - y - 7z = 4$

$6y + 27z = -11$

$0y + 0z = -1$ an inconsistency: so there is no solution.

11. $2x + y + 3z = 8$
$-x + y + z = 10$
$x + y + z = 12$ Interchange the first and third equations, obtaining

$x + y + z = 12$
$-x + y + z = 10$ Add the first equation to the second equation, and add -2 times
$2x + y + 3z = 8$ the first equation to the third equation, obtaining

$x + y + z = 12$
$2y + 2z = 22$
$-y + z = -16$ Interchange the second and third equations, obtaining

$x + y + z = 12$
$-y + z = -16$ Add 2 times the second equation to the third equation,
$2y + 2z = 22$ obtaining

$x + y + z = 12$
$-y + z = -16$
$4z = -10$

From this $z = -5/2$ and back substitution yields $y = 27/2$, $x = 1$.

14. $2x - y + z = 2$
$3x + y + 2z = 3$
$x + y - z = -1$ Interchange the first and third equations, obtaining

$x + y - z = -1$
$3x + y + 2z = 3$ Add -3 times the first equation to the second equation, and add
$2x - y + z = 2$ -2 times the first equation to the third equation, obtaining

$x + y - z = -1$
$-2y + 5z = 6$ Add -1 times the third equation to the second equation,
$-3y + 3z = 4$ obtaining

$x + y - z = -1$
$y + 2z = 2$ Add 3 times the second equation to the third equation,
$-3y + 3z = 4$ obtaining

$x + y - z = -1$
$y + 2z = 2$
$9z = 10$

$z = 10/9$, $y = -2/9$, $x = 1/3$ by substitution

17. $x - 2y + z = -5$

$2x + z = -10$ Add -2 times the first equation to the second equation,

$y - z = 15$ obtaining

$x - 2y + z = -5$

$4y - z = 0$

$y - z = 15$ Interchange the second and third equations, obtaining

$x - 2y + z = -5$

$y - z = 15$ Add -4 times the second equation to the third equation,

$4y - z = 0$ obtaining

$x - 2y + z = -5$

$y - z = 15$

$3z = -60$ From this $\underline{z = -20}$ and $\underline{x = 5}$, $\underline{y = -5}$ by back substituting.

20. Let x = number of chairs, y = number of coffee tables, and z = number of dining tables. Then

$10x + 12y + 15z = 780$ sanding time equation

$6x + 8y + 12z = 540$ staining time equation

$12x + 12y + 18z = 900$ varnishing time equation

Divide the second equation by 2 and the third equation by 6 to simplify

$10x + 12y + 15z = 780$

$3x + 4y + 6z = 270$ Add -1 times the third equation to the second

$2x + 2y + 3z = 150$ equation, obtaining

$10x + 12y + 15z = 780$

$x + 2y + 3z = 120$

$2x + 2y + 3z = 150$ Interchange the first and second equations, obtaining

$x + 2y + 3z = 120$

$10x + 12y + 15z = 780$ Add -10 times the first equation to the second equation,

$2x + 2y + 3z = 150$ and -2 times the first equation to the third, obtaining

$x + 2y + 3z = 120$

$-8y - 15z = -420$ Interchange the second and third equations. Multiply the

$-2y - 3z = -90$ second equation by -1, obtaining

$x + 2y + 3z = 120$

$2y + 3z = 90$ Add 4 times the second equation to the third equation,

$-8y - 15z = -420$ obtaining

20. Continued

$x + 2y + 3z = 120$ From this $z = 20$ dining tables

$2y + 3z = 90$ $y = 15$ coffee tables

$-3z = -60$ $x = 30$ chairs

x and y obtained by back substitution.

Section 6.5 Systems Involving Non-Linear Equations

BASIC CONCEPTS you should learn from this unit.

1. A system of equations containing either one linear equation and one quadratic equation, or two quadratic equations, can be solved by the substitution method or the elimination method.
2. Considering the graph of a system helps in understanding how many solutions there are.

WHAT YOU SHOULD KNOW HOW TO DO upon completion of this unit.

1. Solve a system of two equations containing either one or two second degree equations.
2. Graph systems of equations containing second degree equations.
3. Translate word problems into systems of equations.

SOLUTIONS to Exercise Set 6.5

2. $x^2 + 4y^2 = 32$ Solve the second equation for $x = -2y$ and substitute into the
$x + 2y = 0$ first equation.

$(-2y)^2 + 4y^2 = 32$; $4y^2 + 4y^2 = 32$, $8y^2 = 32$, $y^2 = 4$

and: $y = +2$; $y = -2$ Substituting these in the second equation gives the
$x = -4$; $x = 4$ matching x values.

Graph: an ellipse and a line. There are two solutions.

5. $x^2 - 2x + y^2 = 3$ Solve the second equation for $y = 4 - 2x$ and substitute
$2x + y = 4$ this in the first equation.

$x^2 - 2x + (4 - 2x)^2 = 3$ or $x^2 - 2x + 16 - 16x + 4x^2 = 3$;

or $5x^2 - 18x + 13 = 0$; factoring gives $(5x - 13)(x - 1) = 0$

and so $x = 13/5$, $x = 1$ Substituting these in the second equation gives the
$y = -6/5$, $y = 2$ y values.

Two solutions.

8. $x^2 - y^2 = 3$

$x^2 + y^2 = 5$ Add the first equation to the second equation, obtaining

$x^2 - y^2 = 3$

$2x^2 = 8$

so $x = +2$ $x = -2$

$y = +1$ $y = +1$

Substituting $x^2 = 4$ in the first equation gives the y values. Also

$x = +2$ $x = -2$

$y = -1$ $y = -1$

Hence, there are four solutions.

11. $2y^2 - x^2 = -1$

$4y^2 + x^2 = 25$ Add the first equation to the second equation, obtaining

$2y^2 - x^2 = -1$

$6y^2 = 24$

so $y = +2$ $y = -2$

$x = 3$ $x = 3$

Substituting these values in the first equation gives the x values.

$y = 2$ $y = -2$

$x = -3$ $x = -3$

Hence, there are four solutions.

14. $16y^2 + 5x^2 = 26$ 4 times the first equation and 5 times the second equation

$25y^2 - 4x^2 = 17$ yields

$64y^2 + 20x^2 = 104$

$125y^2 - 20x^2 = 85$ Now add the first equation to the second equation, obtaining

$64y^2 + 20x^2 = 104$

$189y^2 = 189$

so $y^2 = 1$ and

$y = 1$ $y = -1$ Substituting these values in the first equation gives the

$x = \sqrt{2}$ $x = \sqrt{2}$ x values.

$y = 1$ $y = -1$

$x = -\sqrt{2}$ $x = -\sqrt{2}$

17. $x^2 - 3xy - 2y^2 - 2 = 0$

$$x - y - 2 = 0$$

Solving the second equation for y gives $y = x - 2$. Substituting this into the first equation gives:

$$x^2 - 3x(x - 2) - 2(x - 2)^2 - 2 = 0$$

which simplifies to

$$4x^2 - 14x + 10 = 0 \quad \text{or} \quad 2x^2 - 7x + 5 = 0$$

and factoring yields

$$(2x - 5)(x - 1) = 0$$

so $x = 5/2$, $x = 1$. Substituting in the second equation yields the y values $y = 1/2$, $y = -1$.

20. Let x = the width and y = the length. Then $2x + 2y$ = the perimeter and xy = the area, so the two equations become

$$2x + 2y = 22 \qquad\qquad x + y = 11$$
$$\text{or}$$
$$xy = 30 \qquad\qquad xy = 30$$

and from the first equation we get $x = 11 - y$ and substituting in the second we get

$y(11 - y) = 30$ or $11y - y^2 = 30$ or $y^2 - 11y + 30 = 0$; $(y - 5)(y - 6) = 0$,

$y = 6$, or $y = 5$, and hence $x = 5$, $x = 6$.

One dimension is 5 ft, the other is 6 ft.

Section 6.6 Systems of Linear Inequalities in Two Variables

BASIC CONCEPTS you should learn from this unit.

1. The graph of a linear equation in two variables divides the plane into two half planes.
2. The graph of a linear inequality is simply one of the two half planes referred to above and may or may not include the line separating the half planes.
3. The solution to a system of linear inequalities is the set of all points whose coordinates satisfy all the inequalities.
4. The graph of a system of inequalities is the space or area occupied by the satisfying points, usually indicated by shading.
5. Certain word problems are best solved by setting up a system of inequalities and then solving the system.

WHAT YOU SHOULD KNOW HOW TO DO upon completion of this unit.

1. Graph a single inequality, indicating the solution.
2. Graph a system of inequalities and indicate the solution to the system.
3. Solve word problems that lead to a system of linear inequalities.

SOLUTIONS to Exercise Set 6.6

2. $y \geq x + 3$

Graph the line $y = x + 3$; then choose a test point and shade the appropriate half plane. Using (0, 0) as a test point yields a false statement.

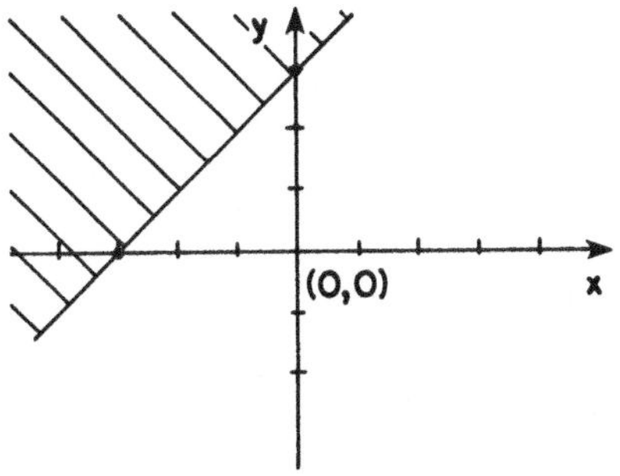

5. $y \leq x - x$

Graph $y = 4 - x$; choose the origin as a test point.

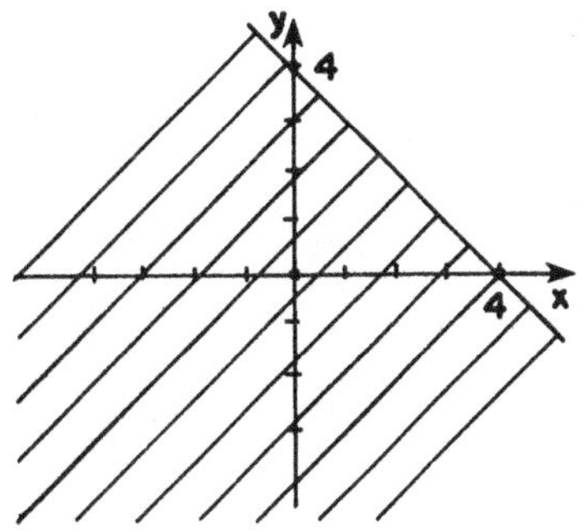

8. $y \leq 2x$

Graph $y = 2x$; choose (1, 1) as a test point.

$1 \leq 2$ which is a true statement.

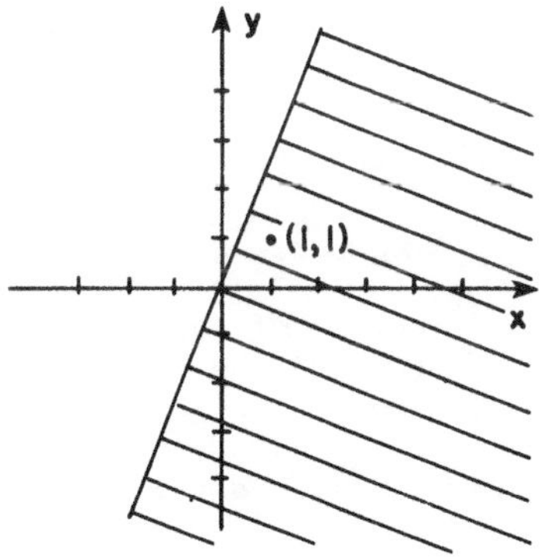

11. $x \leq 4$

Graph $x = 4$;

choose $(0, 0)$ as a test point.

$0 \leq 4$ which is a true statement.

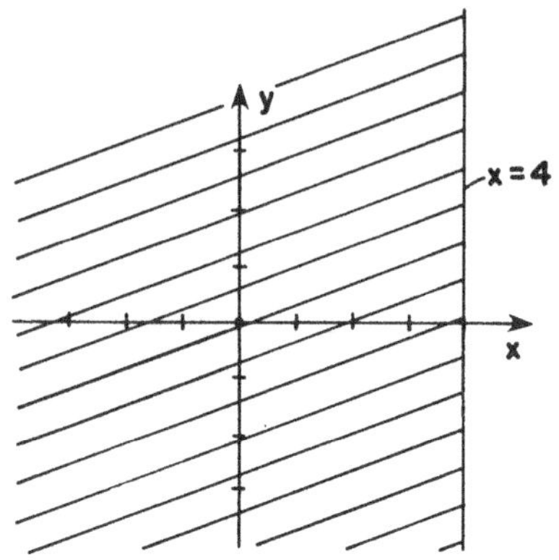

14. $5y \leq 25$

$y \leq 5$

Graph $y = 5$;

choose $(0, 0)$ as a test point.

$0 \leq 5$ which is a true statement.

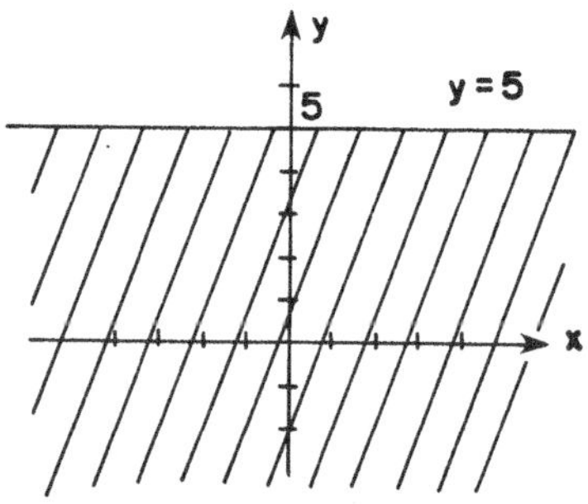

17. $-2 \leq x \leq 3$

Graph $-2 = x$
and $x = 3$.

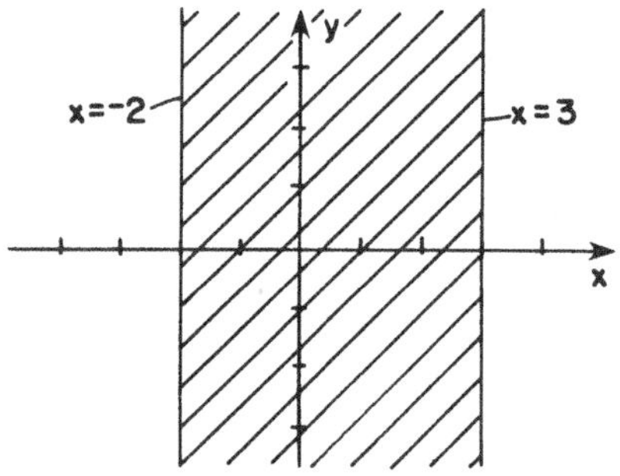

20. Since x = ounces of cheese, $100x$ = number of calories provided by the cheese. Likewise, $110y$ and $80z$ represent the calories in the bread and apples. Then we have

$$100x + 110y + 80z \leq 1500$$

as an appropriate inequality.

23. $3x - y \geq -7$

$3x + y \leq -2$

Graph the lines

$3x - y = -7$ and

$3x + y = -2$.

Try the test points

(0, 0) fails in the second inequality

(0, 6) fails in the second inequality

(-4, 0) fails in the first inequality

(-2, -2) satisfies

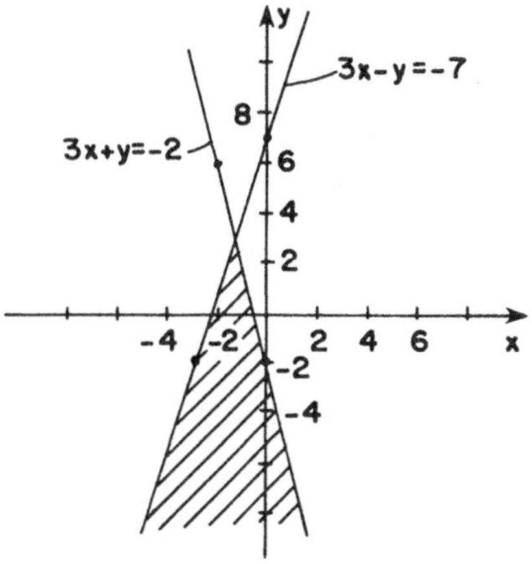

26. $2x - y \geq -3$

$x + y \leq 5$

$x \geq 1$

Graph the lines

$2x - y = -3$

$x + y = 5$

$x = 1$

Try test points to determine the solution.

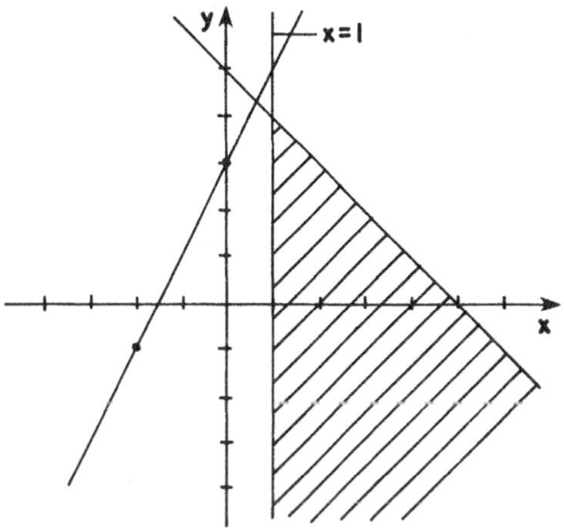

29. $3x + y \leq 6$

$x - 2y \leq -1$

$x \geq 2$

Graph the lines

$3x + y = 6$

$x - 2y = -1$

$x = 2$

Trying test points in all the areas is one way, but noting that satisfying points have to be below the first line, above the second and to the right of the third shows there is no solution.

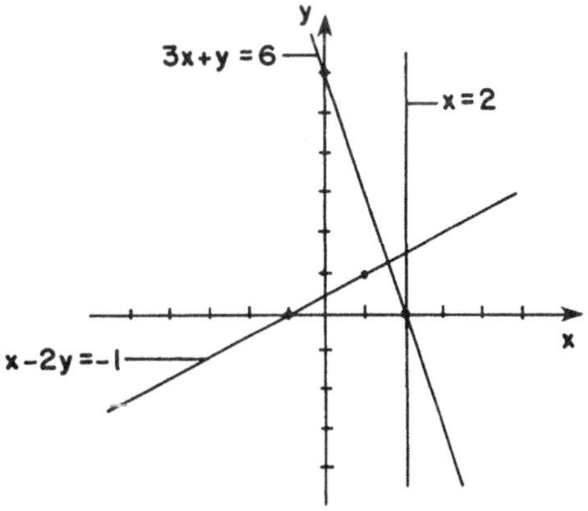

32. $2x + 3y \geq 18$

$x + 3y \geq 12$

$4x + 3y \geq 24$

$x \geq 0$ (y-axis)

$y \geq 0$ (x-axis)

Graphing the lines as before and noting the direction solution points must be from each line yields the solution.

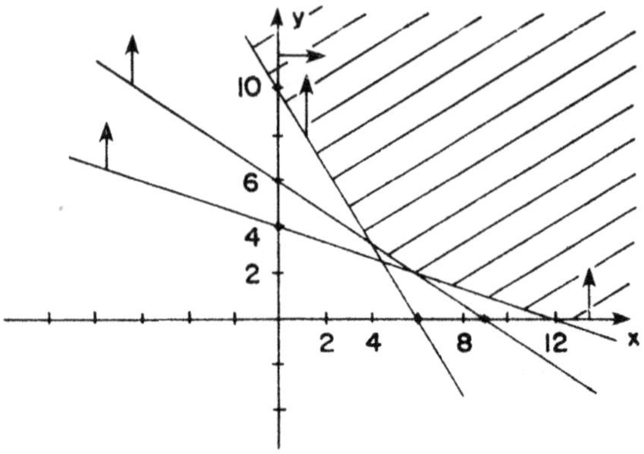

35. Let x = amount in common stocks
y = amount in preferred stocks

Then inequalities can be written:

(a) $x \leq 50,000$

(b) $y \leq 35,000$

(c) $x + y \leq 60,000$

(d) $x \leq 2y$

Also $x \geq 0$

$y \geq 0$

The shaded area represents possible combinations.

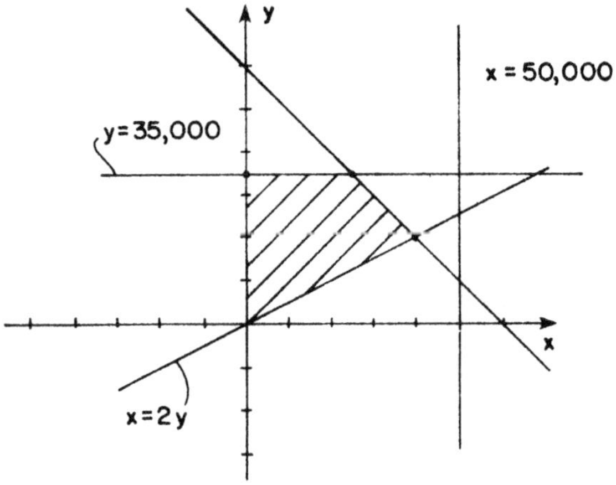

SOLUTIONS to Review Exercises Chapter 6

2. $4x - y = 4$

$2x + y = -1$

Graph each line on the same axes.

Estimate the intersection as $\left(\frac{1}{2}, -2\right)$.

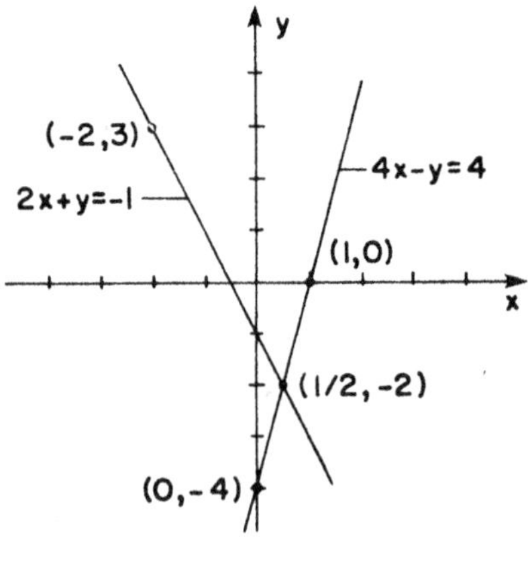

5. $2x + y = 0$

$x - 3y = \frac{7}{4}$

Solving the first equation for y yields $y = -2x$; substitute this for y in the second equation.

$$x - 3(-2x) = \frac{7}{4}; \quad x + 6x = \frac{7}{4}; \quad 7x = \frac{7}{4}, \quad x = \frac{1}{4}$$

and then $y = -\frac{1}{2}$.

8. $2x + 3y = -7$

$-x + 2y = 7$

Solving the second equation for x yields $x = 2y - 7$; substitute this in the first equation.

$$2(2y - 7) + 3y = -7; \quad 4y - 14 + 3y = -7; \quad y = 1, \quad x = -5$$

11. $-3x + y = -13$

$2x - 3y = 11$

Multiply the first equation by 3; then add.

$$\begin{array}{rcr} -9x + 3y &=& -39 \\ 2x - 3y &=& 11 \\ \hline -7x &=& -28 \end{array}$$

and so $x = 4$, $y = -1$ by substituting 4 for x in either equation.

14. $\frac{1}{3}x + \frac{1}{2}y = -1$

$\frac{1}{2}x - \frac{1}{4}y = \frac{5}{2}$

Multiply the second equation by 2, then add.

$$\begin{array}{rcr} \frac{1}{3}x + \frac{1}{2}y &=& -1 \\ x - \frac{1}{2}y &=& 5 \\ \hline \frac{4}{3}x &=& 4 \end{array}$$

Then $x = 3$, $y = -4$.

17. x = cost of hamburger $\qquad 5x + 4y = 22$

y = cost of steak $\qquad 3x + 7y = 28.15$

Multiply the first equation by -3 and the second equation by 5; then add.

$$\begin{array}{rcr} -15x - 12y &=& -66 \\ 15x + 35y &=& 140.75 \\ \hline 23y &=& 74.75 \end{array}$$

$$y = 3.25, \quad x = \$1.80$$

20. $C = 4025 + 9x$

$R = 16x$

Since the break-even point occurs when costs equal revenue, we simply set the right-hand sides of the equations equal to each other:

$$16x = 4025 + 9x$$

and thus $x = 575$ hours and the total revenue at $x = 575$ is $16 \cdot 575 = \$9,200$.

23. $5x - y + 2z = 10$

$-2x + 3y - z = -7$

$3x + 2z = 7$ Add the third equation to the second equation, obtaining

$5x - y + 2z = 10$

$x + 3y + z = 0$

$3x + 2z = 7$ Interchange the first and second equations, obtaining

$x + 3y + z = 0$

$5x - y + 2z = 10$ Add -5 times the first equation to the second equation, and

$3x + 2z = 7$ add -3 times the first equation to the third equation, obtaining

$x + 3y + z = 0$

$-16y - 3z = 10$ Add -9 times the second equation to 16 times the third

$-9y - z = 7$ equation, obtaining

$x + 3y + z = 0$

$-16y - 3z = 10$

$11z = 22$

From this $z = 2$, then $y = -1$, $x = 1$ by back substituting.

26. $x + 2y = 0$

$\underline{-x + 4y = 5}$ Add the equations.

$6y = 5$; $y = 5/6$, then $x = -5/3$.

29. $x^2 + y^2 = 25$

$3x + y = 15$

Solving the second equation for $y = -3x + 15$ and substituting in the first equation yields:

$x^2 + (-3x + 15)^2 = 25$; $x^2 + 9x^2 - 90x + 225 - 25 = 0$;

$10x^2 - 90x + 200 = 0$; $x^2 - 9x + 20 = 0$; $(x - 5)(x - 4) = 0$

$x = 5$ $x = 4$

$x = 0$ $y = 3$

32. $x^2 + y^2 = 9$

$y = x^2 + 3$

From the first $x^2 = 9 - y^2$; substitute this in the second equation to get

$y = 9 - y^2 + 3$ or $y^2 + y - 12 = 0$

and by factoring

$(y + 4)(y - 3) = 0$; $y = -4$ and $y = 3$.

Substituting -4 in the second equation fails to yield <u>real</u> x values. But $y = 3$ yields $x = 0$ for a good solution.

35. $x^2 + y^2 = 16$

$x^2 + y = 1$

Estimate: $x = 2.1$

$y = -3.8$

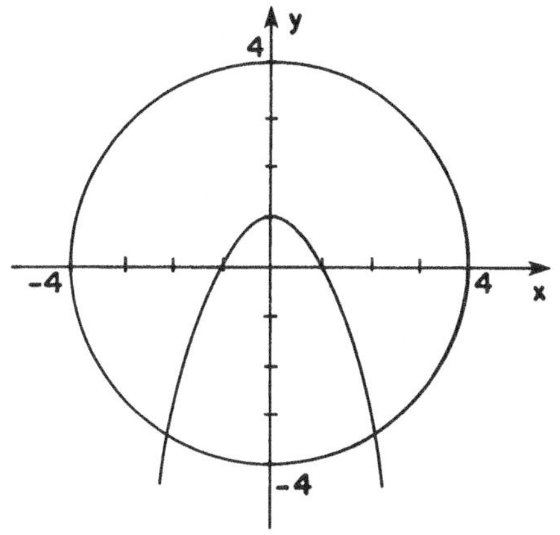

38. $2x + 3y \leq 2$

$x - y \geq 1$

Graph the lines

$2x + 3y = 2$

$x - y = 1$

Note where, in relation to the lines, the solution points must lie (below both lines).

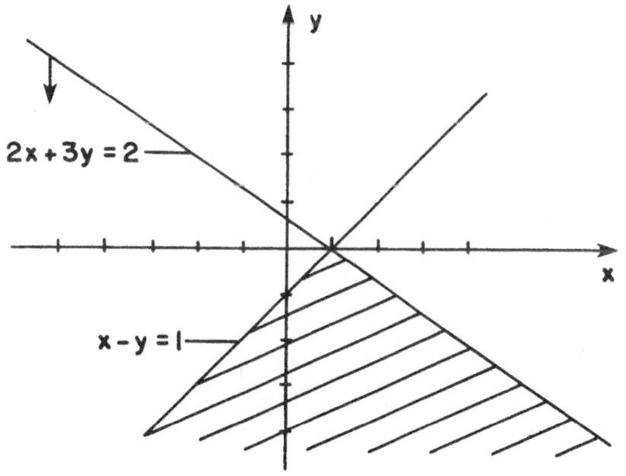

41. $2x + y \leq 4$

$2x - y \leq 3$

$y \geq 0$ (x-axis)

$x \geq 0$ (y-axis)

Graph the lines

$2x + y = 4$

$2x - y = 3$

Note that the solution is below the first line <u>and</u> above the second line, and above the x-axis and to the right of the y-axis.

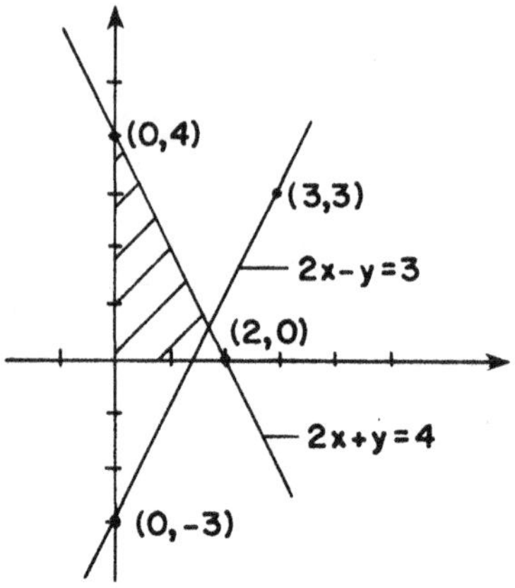

Chapter 6 - A SHORT QUIZ

1. What is meant when we say an (x, y) pair of numbers is a solution to a system of equations?
2. If a system of equations does <u>not</u> have a solution, the system is said to be ___?___
3. If a system of equations has one or more solutions, it is a ___?___ system.
4. Is (-2, 4) a solution to the system

   $x - 2y = -10$

   $2x + 3y = 8$?

   Show work to justify your answer.
5. Draw sketches to illustrate all the possibilities for a system of two linear equations.
6. In solving the system in question 4 by substitution the first step might be to solve the first equation for x, getting $x = 2y - 10$. What is the next step?
7. If a system of two linear equations has exactly one solution, where on the graph of the system does the solution appear?
8. If the two equations of a system are really just different forms of the same equation, how many solutions are there?
9. If in solving a system by elimination $0x + 0y = 0$ is obtained, what is signified?
10. To solve the system

    $2x - 3y = 5$

    $5x + 9y = 4$

    by elimination of the y variable, what is your first step?
11. Sometimes $0x + 0y = k$ $(k \neq 0)$ is obtained. Example: $0 = 7$. Since this is ___?___ true the system is ___?___ .

12. In a business situation of manufacturing and selling a product the <u>equilibrium price</u> is the price of the product at which ___?___ and ___?___ are equal.

13. In solving a system of three linear equations by Gaussian elimination the operations allowed are (a) ___?___ any two equations; (b) ___?___ an equation by a nonzero constant; and (c) ___?___ an equation by ___?___ .

14. When the Gaussian elimination is completed the value of one variable is evident. How are the values of the other variables found?

15. In this system what is the maximum number of solutions you could expect? Why?

$$x^2 + y^2 = 10$$
$$x + y = -3$$

16. If a system of equations consists of an equation of an ellipse and an equation of a parabola, and if the system has three solutions, sketch a graph of this situation.

17. The solution to a linear inequality is generally a ___?___ .

18. Is the solution to $y > 3x - 1$ above or below the line $y = 3x - 1$?

19. Does the test point (3, 0) satisfy this system?

$$2x - 3y \geq 2$$
$$x + y \leq 6$$

20. Try to graph the solution set to this system.

$$y \geq x^2$$
$$x + y \leq 3$$

Quiz Answers Chapter 6

1. When substituted into each equation a true statement results.

2. inconsistent

3. consistent

4. <u>yes</u> $-2 - 8 = -10$
   $-4 + 12 = 8$

5.

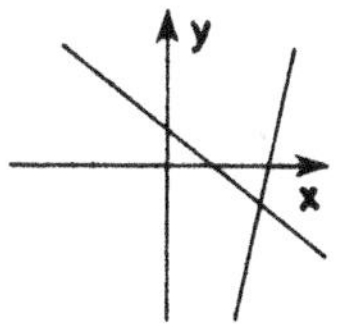

one solution

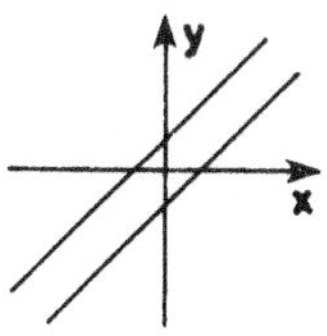

no solution

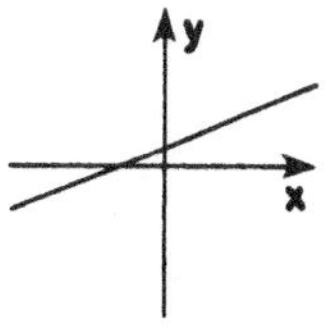

same line

6. Substitute $(2y - 10)$ for x in the second equation.

7. at the intersection of the lines

8. infinitely many

9. infinitely many solutions

10. Multiply the first equation by 3 .

11. never; inconsistent

12. supply, demand

13. (a) interchange; (b) multiply; (c) replace, the sum of itself and a nonzero multiple of another equation.

14. by back substituting

15. Two. The system is a line and a circle.

16.

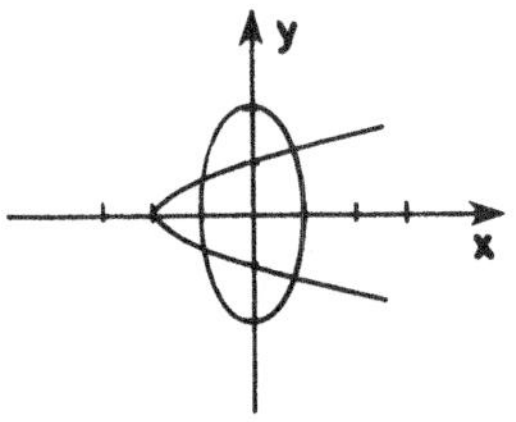

17. a half plane

18. below

19. no ($6^6 - 0 \leq 2$ is false)

20.

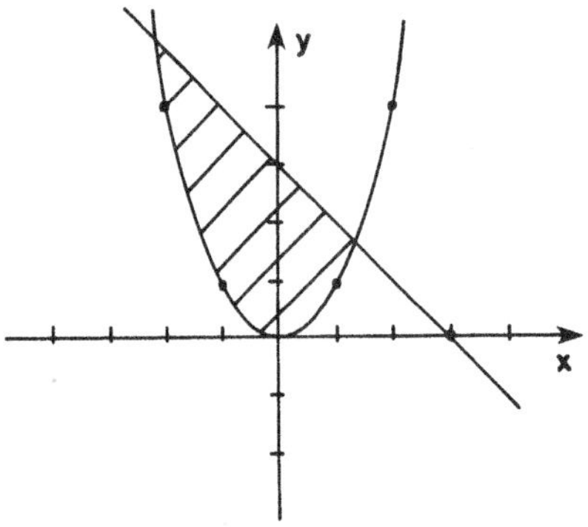

Try some test points.

CHAPTER 7

Matrices and Determinants

Section 7.1 Matrices and Linear Systems

BASIC CONCEPTS you should learn from this unit.

1. Definition of a matrix as a rectangular array of numbers.

2. By conventional agreement, an $m \times n$ matrix has m rows and n columns.

3. A square matrix has as many columns as rows.

4. Double subscripting is a convenient way to identify elements in a matrix. a_{ij} is the element in the ith row and jth column.

5. From a system of linear equations we obtain a coefficient matrix using the coefficients of the variables as entries, and an augmented matrix, which includes the right-hand sides of the equations.

6. The methods of Gaussian elimination - now called elementary row operations - can be applied to an augmented matrix of a linear system of equations to obtain the solution.

7. Obtaining pivot elements (a_{11}, a_{22}, etc.) equal to 1 is helpful.

8. The Gauss-Jordan variation simply has zeros above and below the pivot elements that are 1. The solution then appears in the right column.

WHAT YOU SHOULD KNOW HOW TO DO upon completion of this unit.

1. Define "matrix."

2. Write the coefficient matrix and the augmented matrix for a system of linear equations.

3. Identify elements of a matrix by double subscripting.

4. Solve a system of linear equations by applying row operations to the augmented matrix of the system.

5. Solve a system of linear equations by Gauss-Jordan elimination.

SOLUTIONS to Exercise Set 7.1

2. 1×4 (one row, four columns)

5. 3×3

8. $b_{13} = 8$, $b_{21} = 4$, $b_{33} = -6$, $b_{42} = 9$

11. (a) $\begin{bmatrix} \frac{1}{2} & 1 & 1 \\ 2 & -1 & -4 \\ 4 & 2 & -3 \end{bmatrix}$ (b) $\begin{bmatrix} \frac{1}{2} & 1 & 1 & 4 \\ 2 & -1 & -4 & 6 \\ 4 & 2 & -3 & 8 \end{bmatrix}$

14. $\begin{aligned} 4x \quad\quad &= 2 \\ -7x + 8y &= 3 \end{aligned}$

17. $\begin{bmatrix} 1 & 2 & 0 & 3 \\ 0 & 1 & -2 & 4 \\ 0 & 0 & 1 & 2 \end{bmatrix}$ From the third row we simply read that $\underline{z = 2}$.

Now back substitute in row 2: $y - 2(2) = 4$, so that $\underline{y = 8}$; and finally substitute the z and y values in row 1: $x + 2(8) + 0 = 3$; so $\underline{x = -13}$.

20. $\begin{bmatrix} 1 & -4 & 2 & -4 \\ 0 & 1 & 3 & -2 \\ 0 & 0 & 1 & 5 \end{bmatrix}$ From the last row we get $z = 5$ and back substitution gives $y = -17$ and $z = -82$.

23. The augmented matrix:

$$\begin{bmatrix} 1 & 1 & 1 & 4 \\ 2 & -1 & 2 & 11 \\ 1 & 2 & 1 & 3 \end{bmatrix}, \begin{bmatrix} 1 & 1 & 1 & 4 \\ 0 & -3 & 0 & 3 \\ 0 & 1 & 0 & -1 \end{bmatrix}, \begin{bmatrix} 1 & 1 & 1 & 4 \\ 0 & 1 & 0 & -1 \\ 0 & 1 & 0 & -1 \end{bmatrix},$$

$$\begin{bmatrix} 1 & 1 & 1 & 4 \\ 0 & 1 & 0 & -1 \\ 0 & 0 & 0 & 0 \end{bmatrix}$$

which indicates many solutions with $y = -1$. We may choose z at random . One solution is $(2, -1, 3)$; another is $(0, -1, 5)$.

26. $$\begin{bmatrix} 2 & 1 & -1 & -2 \\ -2 & -2 & 3 & 2 \\ 3 & 1 & -1 & -4 \end{bmatrix}, \begin{bmatrix} 2 & 1 & -1 & -2 \\ -2 & -2 & 3 & 2 \\ 1 & -1 & 2 & -2 \end{bmatrix}, \begin{bmatrix} 1 & -1 & 2 & -2 \\ -2 & -2 & 3 & 2 \\ 0 & -1 & 2 & 0 \end{bmatrix},$$

$$\begin{bmatrix} 1 & -1 & 2 & -2 \\ 0 & -4 & 7 & -2 \\ 0 & +1 & -2 & 0 \end{bmatrix}, \begin{bmatrix} 1 & -1 & 2 & -2 \\ 0 & 1 & -2 & 0 \\ 0 & -4 & 7 & -2 \end{bmatrix}, \begin{bmatrix} 1 & -1 & 2 & -2 \\ 0 & 1 & -2 & 0 \\ 0 & 0 & -1 & -2 \end{bmatrix}$$

And $z = 2$; Substituting in the second row gives

$y = 4$; and these two values in row 1 give

$x = -2$.

29. $$\begin{bmatrix} 1 & 1 & -1 & 2 & 0 \\ 2 & 1 & 0 & -1 & -2 \\ 3 & 0 & 2 & 0 & -3 \\ -1 & 2 & 0 & 3 & 1 \end{bmatrix}, \begin{bmatrix} 1 & 1 & -1 & 2 & 0 \\ 0 & -1 & 2 & -5 & -2 \\ 0 & -3 & 5 & -6 & -3 \\ 0 & 3 & -1 & 5 & 1 \end{bmatrix}, \begin{bmatrix} 1 & 1 & -1 & 2 & 0 \\ 0 & 1 & -2 & 5 & 2 \\ 0 & -3 & 5 & -6 & -3 \\ 0 & 3 & -1 & 5 & 1 \end{bmatrix},$$

$$\begin{bmatrix} 1 & 1 & -1 & 2 & 0 \\ 0 & 1 & -2 & 5 & 2 \\ 0 & 0 & -1 & 9 & 3 \\ 0 & 0 & 5 & -10 & -5 \end{bmatrix}, \begin{bmatrix} 1 & 1 & -1 & 2 & 0 \\ 0 & 1 & -2 & 5 & 2 \\ 0 & 0 & 1 & -9 & -3 \\ 0 & 0 & 1 & -2 & -1 \end{bmatrix}, \begin{bmatrix} 1 & 1 & -1 & 2 & 0 \\ 0 & 1 & -2 & 5 & 2 \\ 0 & 0 & 1 & 9 & -3 \\ 0 & 0 & 0 & 7 & 2 \end{bmatrix}$$

From the last row, now we get $W = 2/7$

and this back substituted yields $z = -3/7$

and these two values in the second row yield $y = -2/7$

and these three put in the first row yield $x = -5/7$

32. $$\begin{bmatrix} 2 & 1 & -1 \\ 3 & -1 & -7 \end{bmatrix}, \begin{bmatrix} 2 & 1 & -1 \\ 1 & -2 & -6 \end{bmatrix}, \begin{bmatrix} 1 & -2 & -6 \\ 2 & 1 & -1 \end{bmatrix},$$

$$\begin{bmatrix} 1 & -2 & -6 \\ 0 & 5 & 11 \end{bmatrix}, \begin{bmatrix} 1 & -2 & -6 \\ 0 & 1 & \frac{11}{5} \end{bmatrix}, \begin{bmatrix} 1 & 0 & -\frac{8}{5} \\ 0 & 1 & \frac{11}{5} \end{bmatrix},$$

which gives directly $x = -8/5$, $y = 11/5$.

35. $$\begin{bmatrix} 2 & 1 & -1 & 9 \\ 1 & -2 & 2 & -3 \\ 3 & 3 & 4 & 11 \end{bmatrix}, \begin{bmatrix} 1 & -2 & 2 & -3 \\ 2 & 1 & -1 & 9 \\ 3 & 3 & 4 & 11 \end{bmatrix}, \begin{bmatrix} 1 & -2 & 2 & -3 \\ 0 & 5 & -5 & 15 \\ 0 & 9 & -2 & 20 \end{bmatrix},$$

$$\begin{bmatrix} 1 & -2 & 2 & -3 \\ 0 & 1 & -1 & 3 \\ 0 & 9 & -2 & 20 \end{bmatrix}, \begin{bmatrix} 1 & -2 & 2 & -3 \\ 0 & 1 & -1 & 3 \\ 0 & 0 & 7 & -7 \end{bmatrix}, \begin{bmatrix} 1 & -2 & 2 & -3 \\ 0 & 1 & -1 & 3 \\ 0 & 0 & 1 & -1 \end{bmatrix},$$

$$\begin{bmatrix} 1 & -2 & 2 & -3 \\ 0 & 1 & 0 & 2 \\ 0 & 0 & 1 & -1 \end{bmatrix}, \begin{bmatrix} 1 & -2 & 0 & -1 \\ 0 & 1 & 0 & 2 \\ 0 & 0 & 1 & -1 \end{bmatrix}, \begin{bmatrix} 1 & 0 & 0 & 3 \\ 0 & 1 & 0 & 2 \\ 0 & 0 & 1 & -1 \end{bmatrix},$$

from which now we read $x = 3$, $y = 2$, $z = -1$.

38. $\begin{bmatrix} 4 & 1 & -1 & -1 \\ 1 & -1 & 2 & 3 \\ -1 & 2 & -1 & 0 \end{bmatrix}$, $\begin{bmatrix} 1 & -1 & 2 & 3 \\ 4 & 1 & -1 & -1 \\ -1 & 2 & -1 & 0 \end{bmatrix}$, $\begin{bmatrix} 1 & -1 & 2 & 3 \\ 0 & 5 & -9 & -13 \\ 0 & 1 & 1 & 3 \end{bmatrix}$,

$\begin{bmatrix} 1 & -1 & 2 & 3 \\ 0 & 1 & 1 & 3 \\ 0 & 5 & -9 & -13 \end{bmatrix}$, $\begin{bmatrix} 1 & -1 & 2 & 3 \\ 0 & 1 & 1 & 3 \\ 0 & 0 & -14 & -28 \end{bmatrix}$, $\begin{bmatrix} 1 & -1 & 2 & 3 \\ 0 & 1 & 1 & 3 \\ 0 & 0 & 1 & 2 \end{bmatrix}$,

$\begin{bmatrix} 1 & -1 & 2 & 3 \\ 0 & 1 & 0 & 1 \\ 0 & 0 & 1 & 2 \end{bmatrix}$, $\begin{bmatrix} 1 & -1 & 0 & -1 \\ 0 & 1 & 0 & 1 \\ 0 & 0 & 1 & 2 \end{bmatrix}$, $\begin{bmatrix} 1 & 0 & 0 & 0 \\ 0 & 1 & 0 & 1 \\ 0 & 0 & 1 & 2 \end{bmatrix}$

so the solution is $x = 0$, $y = 1$, $z = 2$.

Section 7.2 Matrix Operations and Applications

BASIC CONCEPTS you should learn from this unit.

1. Definition of equal matrices: Two matrices are equal if they have the same dimension and their corresponding elements are equal.

2. Matrices are added by adding corresponding elements.

3. Definition of scalar multiplication: Multiply each element of the matrix by the scalar.

4. Definition of matrix multiplication.

5. The product of an $m \times n$ matrix times an $n \times r$ matrix is a matrix of dimension $m \times r$.

6. Matrix multiplication <u>is not</u> commutative.

7. Matrix multiplication <u>is</u> associative.

WHAT YOU SHOULD KNOW HOW TO DO upon completion of this unit.

1. Define equal matrices.

2. Add two matrices.

3. Multiply a matrix by a scalar.

4. Recognize when two matrices can and cannot be multiplied.

5. Multiply two matrices.

6. Write a matrix equation for a linear system of equations.

SOLUTIONS to Exercise Set 7.2

2. Since corresponding elements must be equal

$a = 5$, $b = -6$ (since $a + b = -1$), $c = 3$ (since $2c = 6$),

$d = -7$ (since $c - d = 10$)

5. $$2A = 2\begin{bmatrix} 2 & 3 & 1 \\ -3 & 4 & 1 \end{bmatrix} = \begin{bmatrix} 4 & 6 & 2 \\ -6 & 8 & 2 \end{bmatrix} \quad \text{and}$$

$$3G = 3\begin{bmatrix} -2 & 4 & 2 \\ 1 & 0 & 3 \end{bmatrix} = \begin{bmatrix} -6 & 12 & 6 \\ 3 & 0 & 9 \end{bmatrix}, \quad \text{then}$$

adding corresponding elements yields the sum

$$\begin{bmatrix} -2 & 18 & 8 \\ -3 & 8 & 11 \end{bmatrix}$$

8. $$2B = 2\begin{bmatrix} 2 & -1 \\ 3 & 2 \\ 4 & 1 \end{bmatrix} = \begin{bmatrix} 4 & -2 \\ 6 & 4 \\ 8 & 2 \end{bmatrix}, \quad \text{a } 3 \times 2 \text{ matrix}$$

and since D is a 2×2 matrix, $2B - D$ is <u>not possible</u>.

11. The product CB will have dimension 3×2, which cannot be added to D, a 2×2 matrix.

14. $$\begin{bmatrix} 2 & 3 & 1 \\ -3 & 4 & 1 \end{bmatrix} \cdot \begin{bmatrix} 1 & 2 & 3 \\ 4 & -1 & 2 \\ 3 & 2 & 5 \end{bmatrix} = \begin{bmatrix} 17 & 3 & 17 \\ 16 & -8 & 4 \end{bmatrix} = AC$$

$$2 \cdot \begin{bmatrix} -3 & 2 \\ 4 & 1 \end{bmatrix} \cdot \begin{bmatrix} -2 & 4 & 2 \\ 1 & 0 & 3 \end{bmatrix} = \begin{bmatrix} -6 & 4 \\ 8 & 2 \end{bmatrix} \cdot \begin{bmatrix} -2 & 4 & 2 \\ 1 & 0 & 3 \end{bmatrix} = \begin{bmatrix} 16 & -24 & 0 \\ -14 & 32 & 22 \end{bmatrix} = 2DC$$

Adding corresponding elements yields

$$\begin{bmatrix} 33 & -21 & 17 \\ 2 & 24 & 26 \end{bmatrix}$$

17.
$$2GE = 2\begin{bmatrix} -2 & 4 & 2 \\ 1 & 0 & 3 \end{bmatrix} \cdot \begin{bmatrix} 1 & -3 & 2 \\ 3 & 2 & 4 \\ 1 & 1 & 2 \end{bmatrix}$$

$$= \begin{bmatrix} -4 & 8 & 4 \\ 2 & 0 & 6 \end{bmatrix} \cdot \begin{bmatrix} 1 & -3 & 2 \\ 3 & 2 & 4 \\ 1 & 1 & 2 \end{bmatrix} = \begin{bmatrix} 24 & 32 & 32 \\ 8 & 0 & 16 \end{bmatrix}$$

and

$$3A = \begin{bmatrix} 6 & 9 & 3 \\ -9 & 12 & 3 \end{bmatrix}$$

Now subtracting $2GE - 3A$ yields

$$\begin{bmatrix} 18 & 23 & 29 \\ 17 & -12 & 13 \end{bmatrix}$$

20.
$$AB = \begin{bmatrix} 1 & 2 \\ 3 & 2 \end{bmatrix} \cdot \begin{bmatrix} 2 & -1 \\ -3 & 4 \end{bmatrix} = \begin{bmatrix} -4 & 7 \\ 0 & 5 \end{bmatrix}$$

and

$$BA = \begin{bmatrix} 2 & -1 \\ -3 & 4 \end{bmatrix} \cdot \begin{bmatrix} 1 & 2 \\ 3 & 2 \end{bmatrix} = \begin{bmatrix} -1 & 2 \\ 9 & 2 \end{bmatrix}$$

clearly $\neq AB$.

23. Applying the rules for multiplying, we find

$$AI = \begin{bmatrix} (a_{11} + 0 + 0) & (0 + a_{12} + 0) & (0 + 0 + a_{13}) \\ (a_{21} + 0 + 0) & (0 + a_{22} + 0) & (0 + 0 + a_{23}) \\ (a_{31} + 0 + 0) & (0 + a_{32} + 0) & (0 + 0 + a_{33}) \end{bmatrix} = \begin{bmatrix} a_{11} & a_{12} & a_{13} \\ a_{21} & a_{22} & a_{23} \\ a_{31} & a_{32} & a_{33} \end{bmatrix} = A$$

Also

$$IA = \begin{bmatrix} (a_{11} + 0 + 0) & (a_{12} + 0 + 0) & (a_{13} + 0 + 0) \\ (0 + a_{21} + 0) & (0 + a_{22} + 0) & (0 + a_{23} + 0) \\ (0 + 0 + a_{31}) & (0 + 0 + a_{32}) & (0 + 0 + a_{33}) \end{bmatrix} = \begin{bmatrix} a_{11} & a_{12} & a_{13} \\ a_{21} & a_{22} & a_{23} \\ a_{31} & a_{32} & a_{33} \end{bmatrix} = A$$

26. Let A be the coefficient matrix of the system, X be a column matrix $\begin{bmatrix} x \\ y \end{bmatrix}$, and B be a column matrix $\begin{bmatrix} 6 \\ -2 \end{bmatrix}$. Then $AX = B$ is

$$\begin{bmatrix} 7 & -2 \\ -2 & 3 \end{bmatrix} \cdot \begin{bmatrix} x \\ y \end{bmatrix} = \begin{bmatrix} 6 \\ -2 \end{bmatrix}$$

29. Following a procedure similar to that shown in solution 26, we get

$$AX = B: \quad \begin{bmatrix} 3 & -1 & 4 \\ 2 & 2 & \frac{3}{4} \\ 1 & -\frac{1}{4} & 1 \end{bmatrix} \cdot \begin{bmatrix} x \\ y \\ z \end{bmatrix} = \begin{bmatrix} 5 \\ -1 \\ \frac{1}{2} \end{bmatrix}$$

32. Finding AX and setting it equal to B, we have

$$\begin{bmatrix} 1 & 7 & -2 \\ 3 & 6 & 1 \\ -4 & 2 & 0 \end{bmatrix} \cdot \begin{bmatrix} x \\ y \\ z \end{bmatrix} = \begin{bmatrix} 3 \\ -3 \\ 2 \end{bmatrix}$$

or finally

$$\begin{aligned} x + 7y - 2z &= 3 \\ 3x + 6y + z &= -3 \\ -4x + 2y &= 2 \end{aligned}$$

as the system.

35.

$$AI_n = \begin{bmatrix} a_{11} & a_{12} \cdots & a_{1n} \\ a_{21} & a_{22} \cdots & a_{2n} \\ \vdots & & \vdots \\ a_{n1} & \cdots & a_{nn} \end{bmatrix} \cdot \begin{bmatrix} 1 & 0 \ldots & 0 \\ 0 & 1 \ldots & 0 \\ \vdots & & \vdots \\ 0 & \ldots & 1 \end{bmatrix} = \begin{bmatrix} a_{11} & a_{12} \cdots & a_{1n} \\ \vdots & & \vdots \\ a_{n1} & \cdots & a_{nn} \end{bmatrix}$$

To see this notice that the element in the first row, first column of the product is really the sum of $a_{11} \cdot 1 + (n - 1)$ zeros or simply a_{11}. Also the element in the first row, second column of the product is $0 + a_{12} + (n - 2)$ zeros. Similar arguments can be made for each element in the product matrix, which turns out to be A itself.

$$I_nA = \begin{bmatrix} 1 & 0 \ldots & 0 \\ 0 & 1 \ldots & 0 \\ \vdots & & \vdots \\ 0 & \ldots & 1 \end{bmatrix} \cdot \begin{bmatrix} a_{11} & a_{12} \cdots & a_{1n} \\ a_{21} & a_{22} \cdots & a_{2n} \\ \vdots & & \vdots \\ a_{n1} & \cdots & a_{nn} \end{bmatrix} = \begin{bmatrix} a_{11} & a_{12} \cdots & a_{1n} \\ a_{21} & \cdots & \vdots \\ \vdots & & \\ a_{n1} & \cdots & a_{nn} \end{bmatrix}$$

which is also equal to A, by a similar discussion. Now since $AI_n = A$ and $IA_n = A$, $AI_n = I_nA$, as required.

BASIC CONCEPTS you should learn from this unit.

1. A square $n \times n$ matrix A is invertible if an $n \times n$ matrix B can be found such that $AB = BA = I_n$, I_n being the identity matrix of order n. Then B is the inverse of A.

2. The inverse of matrix A, that is, A^{-1}, can be found by considering the matrix

$$\left[\begin{array}{c|c} A & I_n \end{array}\right]$$

and by row operations converting A into I_n. Then I_n (the right portion) will become A^{-1}.

3. Not all $n \times n$ matrices are invertible and the process of concept 2 will reveal which are or are not.

4. If $AX = B$ is a linear system, A being the coefficient matrix and if the inverse A^{-1}, of A exists, then the system has exactly one solution

$$X = A^{-1}B$$

5. Use of the inverse of a matrix to solve systems is especially useful if we have several systems with the same coefficient matrix.

WHAT YOU SHOULD KNOW HOW TO DO upon completion of this unit.

1. Verify that a matrix is the inverse of another matrix.

2. Find the inverse of a matrix if it has an inverse.

3. Determine that a matrix does not have an inverse.

4. Solve a system of linear equations using the inverse of the coefficient matrix.

SOLUTIONS to Exercise Set 7.3

2. $$AB = \begin{bmatrix} 3 & -1 \\ -2 & 2 \end{bmatrix} \cdot \begin{bmatrix} 1/2 & 1/4 \\ 1/2 & 3/4 \end{bmatrix} = \begin{bmatrix} 3/2 - 1/2 & 3/4 - 3/4 \\ -1 + 1 & -1/2 + 3/2 \end{bmatrix} = \begin{bmatrix} 1 & 0 \\ 0 & 1 \end{bmatrix}$$

Since $AB = I$, yes, B is the inverse of A.

5. First form the augmented matrix

$$\left[\begin{array}{cc|cc} -1 & 5 & 1 & 0 \\ 2 & -4 & 0 & 1 \end{array}\right]$$

and now transform the left-hand 2×2 matrix into I_2 by row operations;

$$\left[\begin{array}{cc|cc} 1 & -5 & -1 & 0 \\ 0 & 6 & 2 & 1 \end{array}\right]; \quad \left[\begin{array}{cc|cc} 1 & -5 & -1 & 0 \\ 0 & 1 & 1/3 & 1/6 \end{array}\right]; \quad \left[\begin{array}{cc|cc} 1 & 0 & 2/3 & 5/6 \\ 0 & 1 & 1/3 & 1/6 \end{array}\right]$$

Now the right-hand 2×2 matrix is the inverse of A.

8. Form the augmented matrix as

$$\left[\begin{array}{ccc|ccc} 2 & 1 & 0 & 1 & 0 & 0 \\ 1 & 1 & 0 & 0 & 1 & 0 \\ 1 & 1 & 1 & 0 & 0 & 1 \end{array}\right]$$

Now perform row operations to transform the left-hand 3×3 into I_3.

$$\left[\begin{array}{ccc|ccc} 1 & 1 & 1 & 0 & 0 & 1 \\ 1 & 1 & 0 & 0 & 1 & 0 \\ 2 & 1 & 0 & 1 & 0 & 0 \end{array}\right], \left[\begin{array}{ccc|ccc} 1 & 1 & 1 & 0 & 0 & 1 \\ 0 & 0 & -1 & 0 & 1 & -1 \\ 0 & -1 & -2 & 1 & 0 & -2 \end{array}\right], \left[\begin{array}{ccc|ccc} 1 & 1 & 1 & 0 & 0 & 1 \\ 0 & 1 & 2 & -1 & 0 & 2 \\ 0 & 0 & 1 & 0 & -1 & 1 \end{array}\right],$$

$$\left[\begin{array}{ccc|ccc} 1 & 1 & 1 & 0 & 0 & 1 \\ 0 & 1 & 0 & -1 & 2 & 0 \\ 0 & 0 & 1 & 0 & -1 & 1 \end{array}\right], \left[\begin{array}{ccc|ccc} 1 & 1 & 0 & 0 & 1 & 0 \\ 0 & 1 & 0 & -1 & 2 & 0 \\ 0 & 0 & 1 & 0 & -1 & 1 \end{array}\right], \left[\begin{array}{ccc|ccc} 1 & 0 & 0 & 1 & -1 & 0 \\ 0 & 1 & 0 & -1 & 2 & 0 \\ 0 & 0 & 1 & 0 & -1 & 1 \end{array}\right]$$

Now the right-hand 3×3 matrix is the inverse of

$$\begin{bmatrix} 2 & 1 & 0 \\ 1 & 1 & 0 \\ 1 & 1 & 1 \end{bmatrix}$$

11. Form the augmented matrix

$$\left[\begin{array}{cc|cc} 1 & 3 & 1 & 0 \\ -1 & 4 & 0 & 1 \end{array}\right]$$

and transform the left-hand 2×2 matrix into I_2

$$\left[\begin{array}{cc|cc} 1 & 3 & 1 & 0 \\ 0 & 7 & 1 & 1 \end{array}\right], \left[\begin{array}{cc|cc} 1 & 3 & 1 & 0 \\ 0 & 1 & \frac{1}{7} & \frac{1}{7} \end{array}\right], \left[\begin{array}{cc|cc} 1 & 0 & \frac{4}{7} & -\frac{3}{7} \\ 0 & 1 & \frac{1}{7} & \frac{1}{7} \end{array}\right]$$

and the right-hand 2×2 matrix is the inverse required.

14. $$\left[\begin{array}{ccc|ccc} 8 & 7 & -1 & 1 & 0 & 0 \\ -5 & -5 & 1 & 0 & 1 & 0 \\ -4 & -4 & 1 & 0 & 0 & 1 \end{array}\right], \left[\begin{array}{ccc|ccc} 8 & 7 & -1 & 1 & 0 & 0 \\ -5 & -5 & 1 & 0 & 1 & 0 \\ 1 & 1 & 0 & 0 & -1 & 1 \end{array}\right], \left[\begin{array}{ccc|ccc} 1 & 1 & 0 & 0 & -1 & 1 \\ -5 & -5 & 1 & 0 & 1 & 0 \\ 8 & 7 & -1 & 1 & 0 & 0 \end{array}\right],$$

$$\left[\begin{array}{ccc|ccc} 1 & 1 & 0 & 0 & -1 & 1 \\ 0 & 0 & 1 & 0 & -4 & 5 \\ 0 & -1 & -1 & 1 & 8 & -8 \end{array}\right], \left[\begin{array}{ccc|ccc} 1 & 1 & 0 & 0 & -1 & 1 \\ 0 & 1 & 1 & -1 & -8 & 8 \\ 0 & 0 & 1 & 0 & -4 & 5 \end{array}\right], \left[\begin{array}{ccc|ccc} 1 & 1 & 0 & 0 & -1 & 1 \\ 0 & 1 & 0 & -1 & -4 & 3 \\ 0 & 0 & 1 & 0 & -4 & 5 \end{array}\right]$$

$$\left[\begin{array}{ccc|ccc} 1 & 0 & 0 & 1 & 3 & -2 \\ 0 & 1 & 0 & -1 & -4 & 3 \\ 0 & 0 & 1 & 0 & -4 & 5 \end{array}\right]$$

and the right-hand 3×3 matrix is the inverse required.

17. $$\begin{bmatrix} 1 & 0 & -1 & 1 & 0 & 0 \\ 2 & 1 & 0 & 0 & 1 & 0 \\ 0 & 1 & 1 & 0 & 0 & 1 \end{bmatrix}, \begin{bmatrix} 1 & 0 & -1 & 1 & 0 & 0 \\ 0 & 1 & 2 & -2 & 1 & 0 \\ 0 & 1 & 1 & 0 & 0 & 1 \end{bmatrix}, \begin{bmatrix} 1 & 0 & -1 & 1 & 0 & 0 \\ 0 & 1 & 2 & -2 & 1 & 0 \\ 0 & 0 & -1 & 2 & -1 & 1 \end{bmatrix}$$

$$\begin{bmatrix} 1 & 0 & -1 & 1 & 0 & 0 \\ 0 & 1 & 2 & -2 & 1 & 0 \\ 0 & 0 & 1 & -2 & 1 & -1 \end{bmatrix} \quad \left[\begin{array}{ccc|ccc} 1 & 0 & 0 & -1 & 1 & -1 \\ 0 & 1 & 0 & 2 & -1 & 2 \\ 0 & 0 & 1 & -2 & 1 & -1 \end{array}\right]$$

and the right-hand 3×3 matrix is the inverse required.

20. The coefficient matrix is

$$\begin{bmatrix} 2 & -3 \\ 3 & 1 \end{bmatrix}$$

We find its inverse:

$$\left[\begin{array}{cc|cc} 2 & -3 & 1 & 0 \\ 3 & 1 & 0 & 1 \end{array}\right], \begin{bmatrix} 2 & -3 & 1 & 0 \\ 1 & 4 & -1 & 1 \end{bmatrix}, \begin{bmatrix} 1 & 4 & -1 & 1 \\ 2 & -3 & 1 & 0 \end{bmatrix}, \begin{bmatrix} 1 & 4 & -1 & 1 \\ 0 & -11 & 3 & -2 \end{bmatrix},$$

$$\begin{bmatrix} 1 & 4 & -1 & 1 \\ 0 & 1 & -\frac{3}{11} & \frac{2}{11} \end{bmatrix}, \left[\begin{array}{cc|cc} 1 & 0 & \frac{1}{11} & \frac{3}{11} \\ 0 & 1 & -\frac{3}{11} & \frac{2}{11} \end{array}\right]$$

Now find the product of the inverse times the constant-term column matrix:

$$\begin{bmatrix} \frac{1}{11} & \frac{3}{11} \\ -\frac{3}{11} & \frac{2}{11} \end{bmatrix} \cdot \begin{bmatrix} -5 \\ -13 \end{bmatrix} = \begin{bmatrix} -4 \\ -1 \end{bmatrix}$$

Thus $x = -4$ and $y = -1$.

23. Find the inverse of the coefficient matrix

$$\left[\begin{array}{ccc|ccc} 2 & -1 & 3 & 1 & 0 & 0 \\ 3 & -1 & 1 & 0 & 1 & 0 \\ 1 & 1 & 1 & 0 & 0 & 1 \end{array}\right], \left[\begin{array}{ccc|ccc} 1 & 1 & 1 & 0 & 0 & 1 \\ 3 & -1 & 1 & 0 & 1 & 0 \\ 2 & -1 & 3 & 1 & 0 & 0 \end{array}\right], \left[\begin{array}{ccc|ccc} 1 & 1 & 1 & 0 & 0 & 1 \\ 0 & -4 & -2 & 0 & 1 & -3 \\ 0 & -3 & 1 & 1 & 0 & -2 \end{array}\right],$$

$$\left[\begin{array}{ccc|ccc} 1 & 1 & 1 & 0 & 0 & 1 \\ 0 & 1 & 3 & 1 & -1 & 1 \\ 0 & 0 & 10 & 4 & -3 & 1 \end{array}\right], \qquad \left[\begin{array}{ccc|ccc} 1 & 1 & 1 & 0 & 0 & 0 \\ 0 & 1 & 3 & 1 & -1 & 1 \\ 0 & 0 & 1 & 4/10 & -3/10 & 1/10 \end{array}\right],$$

$$\left[\begin{array}{ccc|ccc} 1 & 1 & 1 & 0 & 0 & 1 \\ 0 & 1 & 0 & -2/10 & -1/10 & 7/10 \\ 0 & 0 & 1 & 4/10 & -3/10 & 1/10 \end{array}\right], \left[\begin{array}{ccc|ccc} 1 & 1 & 0 & -4/10 & 3/10 & 9/10 \\ 0 & 1 & 0 & -2/10 & -1/10 & 7/10 \\ 0 & 0 & 1 & 4/10 & -3/10 & 1/10 \end{array}\right],$$

$$\left[\begin{array}{ccc|ccc} 1 & 0 & 0 & -2/10 & 4/10 & 2/10 \\ 0 & 1 & 0 & -2/10 & -1/10 & 7/10 \\ 0 & 0 & 1 & 4/10 & -3/10 & 1/10 \end{array}\right]$$

Then

$$\begin{bmatrix} -\frac{2}{10} & \frac{4}{10} & \frac{2}{10} \\ -\frac{2}{10} & -\frac{1}{10} & \frac{7}{10} \\ \frac{4}{10} & -\frac{3}{10} & \frac{1}{10} \end{bmatrix} \cdot \begin{bmatrix} -11 \\ -5 \\ -1 \end{bmatrix} = \begin{bmatrix} 0 \\ 2 \\ 3 \end{bmatrix}$$

or

$$x = 0$$
$$y = 2$$
$$z = 3$$

26. The coefficient matrix is $\begin{bmatrix} 2 & 1 \\ 3 & -1 \end{bmatrix}$. We find its inverse:

$$\begin{bmatrix} 2 & 1 & 1 & 0 \\ 3 & -1 & 0 & 1 \end{bmatrix}, \begin{bmatrix} 2 & 1 & 1 & 0 \\ 1 & -2 & -1 & 1 \end{bmatrix}, \begin{bmatrix} 1 & -2 & -1 & 1 \\ 2 & 1 & 1 & 0 \end{bmatrix}, \begin{bmatrix} 1 & -2 & -1 & 1 \\ 0 & 5 & 3 & -2 \end{bmatrix},$$

$$\begin{bmatrix} 1 & -2 & -1 & 1 \\ 0 & 1 & 3/5 & -2/5 \end{bmatrix}, \begin{bmatrix} 1 & 0 & 1/5 & 1/5 \\ 0 & 1 & 3/5 & -2/5 \end{bmatrix}$$

and multiply

26. continued

$$\begin{bmatrix} 1/5 & 1/5 \\ 3/5 & -2/5 \end{bmatrix} \cdot \begin{bmatrix} -1 \\ -7 \end{bmatrix} = \begin{bmatrix} -8/5 \\ 11/5 \end{bmatrix}$$

so

$$x = -8/5$$
$$y = 11/5$$

29. We find the inverse of the coefficient matrix:

$$\begin{bmatrix} 2 & 1 & -1 & 1 & 0 & 0 \\ 1 & -2 & 2 & 0 & 1 & 0 \\ 3 & 3 & 4 & 0 & 0 & 1 \end{bmatrix}, \begin{bmatrix} 1 & -2 & 2 & 0 & 1 & 0 \\ 2 & 1 & -1 & 1 & 0 & 0 \\ 3 & 3 & 4 & 0 & 0 & 1 \end{bmatrix}, \begin{bmatrix} 1 & -2 & 2 & 0 & 1 & 0 \\ 0 & 5 & -5 & 1 & -2 & 0 \\ 0 & 9 & -2 & 0 & -3 & 1 \end{bmatrix},$$

$$\begin{bmatrix} 1 & -2 & 2 & 0 & 1 & 0 \\ 0 & 1 & -1 & 1/5 & -2/5 & 0 \\ 0 & 9 & -2 & 0 & -3 & 1 \end{bmatrix}, \begin{bmatrix} 1 & -2 & 2 & 0 & 1 & 0 \\ 0 & 1 & -1 & 1/5 & -2/5 & 0 \\ 0 & 0 & 7 & -9/5 & 3/5 & 1 \end{bmatrix},$$

$$\begin{bmatrix} 1 & -2 & 2 & 0 & 1 & 0 \\ 0 & 1 & -1 & 1/5 & -2/5 & 0 \\ 0 & 0 & 1 & -9/35 & 3/35 & 5/35 \end{bmatrix}, \begin{bmatrix} 1 & -2 & 2 & 0 & 1 & 0 \\ 0 & 1 & 0 & -2/35 & -11/35 & 5/35 \\ 0 & 0 & 1 & -9/35 & 3/35 & 5/35 \end{bmatrix},$$

$$\begin{bmatrix} 1 & -2 & 0 & 18/35 & 29/35 & -10/35 \\ 0 & 1 & 0 & -2/35 & -11/35 & 5/35 \\ 0 & 0 & 1 & -9/35 & 3/35 & 5/35 \end{bmatrix}, \begin{bmatrix} 1 & 0 & 0 & 14/35 & 7/35 & 0 \\ 0 & 1 & 0 & -2/35 & -11/35 & 5/35 \\ 0 & 0 & 1 & -9/35 & 3/35 & 5/35 \end{bmatrix}$$

Multiplying the inverse by $\begin{bmatrix} 9 \\ -3 \\ 11 \end{bmatrix}$ yields $\begin{bmatrix} 3 \\ 2 \\ -1 \end{bmatrix}$

or

$$x = 3$$
$$y = 2$$
$$z = -1$$

32. We find the inverse of the coefficient matrix:

$$\begin{bmatrix} 4 & 1 & -1 & 1 & 0 & 0 \\ 1 & -1 & 2 & 0 & 1 & 0 \\ -1 & 2 & -1 & 0 & 0 & 1 \end{bmatrix}, \begin{bmatrix} 1 & -1 & 2 & 0 & 1 & 0 \\ 4 & 1 & -1 & 1 & 0 & 1 \\ -1 & 2 & -1 & 0 & 0 & 1 \end{bmatrix}, \begin{bmatrix} 1 & -1 & 2 & 0 & 1 & 0 \\ 0 & 5 & -9 & 1 & -4 & 1 \\ 0 & 1 & 1 & 0 & 1 & 1 \end{bmatrix},$$

32. continued

$$\begin{bmatrix} 1 & -1 & 2 & 0 & 1 & 0 \\ 0 & 1 & 1 & 0 & 1 & 1 \\ 0 & 0 & -14 & 1 & -9 & -4 \end{bmatrix}, \begin{bmatrix} 1 & -1 & 2 & 0 & 1 & 0 \\ 0 & 1 & 1 & 0 & 1 & 1 \\ 0 & 0 & 1 & -1/14 & 9/14 & 4/14 \end{bmatrix},$$

$$\begin{bmatrix} 1 & -1 & 0 & 2/14 & -4/14 & -8/14 \\ 0 & 1 & 0 & 1/14 & 5/14 & 9/14 \\ 0 & 0 & 1 & -1/14 & 9/14 & 4/14 \end{bmatrix}, \begin{bmatrix} 1 & 0 & 0 & 3/14 & 1/14 & 1/14 \\ 0 & 1 & 0 & 1/14 & 5/14 & 9/14 \\ 0 & 0 & 1 & -1/14 & 9/14 & 4/14 \end{bmatrix}$$

and multiplying the inverse by $\begin{bmatrix} -1 \\ 3 \\ 0 \end{bmatrix}$ yields $\begin{bmatrix} 0 \\ 1 \\ 2 \end{bmatrix}$

or

$$x = 0$$
$$y = 1$$
$$z = 2$$

35. $X = A^{-1}B_1$ or

$$\begin{bmatrix} 3 & -2 & 4 \\ 2 & -1 & 0 \\ 0 & 4 & 1 \end{bmatrix} \cdot \begin{bmatrix} 1 \\ -1 \\ 5 \end{bmatrix} = \begin{bmatrix} 25 \\ 3 \\ 1 \end{bmatrix}$$

and $X = A^{-1} \cdot B_2$ or

$$\begin{bmatrix} 3 & -2 & 4 \\ 2 & -1 & 0 \\ 0 & 4 & 1 \end{bmatrix} \cdot \begin{bmatrix} 4 \\ 3 \\ -2 \end{bmatrix} = \begin{bmatrix} 2 \\ 5 \\ 10 \end{bmatrix}$$

38. Using x and y as the amounts invested at 10% and 15%, respectively, make three systems:

$$\begin{aligned} x + y &= 3{,}000 \\ 10x + 15y &= 360{,}000 \end{aligned} \; ; \quad \begin{aligned} x + y &= 3{,}000 \\ 10x + 15y &= 400{,}000 \end{aligned} \; ; \quad \begin{aligned} x + y &= 3{,}000 \\ 10x + 15y &= 5{,}000 \end{aligned}$$

Find the inverse of the coefficient matrix:

$$\begin{bmatrix} 1 & 1 & 1 & 0 \\ 10 & 15 & 0 & 1 \end{bmatrix}, \begin{bmatrix} 1 & 1 & 1 & 0 \\ 0 & 5 & -10 & 1 \end{bmatrix}, \begin{bmatrix} 1 & 1 & 1 & 0 \\ 0 & 1 & -2 & 1/5 \end{bmatrix}, \left[\begin{array}{cc|cc} 1 & 0 & 3 & -1/5 \\ 0 & 1 & -2 & 1/5 \end{array}\right]$$

Multiply the inverse times each constant-term column matrix:

(a) $\begin{bmatrix} 3 & -1/5 \\ -2 & 1/5 \end{bmatrix} \cdot \begin{bmatrix} 30{,}000 \\ 360{,}000 \end{bmatrix} = \begin{bmatrix} \$18{,}000 \\ \$12{,}000 \end{bmatrix} \begin{matrix} = x \\ = y \end{matrix}$

38. continued

(b) $\begin{bmatrix} 3 & -1/5 \\ -2 & 1/5 \end{bmatrix} \cdot \begin{bmatrix} 30{,}000 \\ 400{,}000 \end{bmatrix} = \begin{bmatrix} \$10{,}000 \\ \$20{,}000 \end{bmatrix} \begin{matrix} = x \\ = y \end{matrix}$

(c) $\begin{bmatrix} 3 & -1/5 \\ -2 & 1/5 \end{bmatrix} \cdot \begin{bmatrix} 30{,}000 \\ 500{,}000 \end{bmatrix} = \begin{bmatrix} -10{,}000 \\ 40{,}000 \end{bmatrix}$

which is not possible!

Section 7.4 Determinants and Cramer's Rule

BASIC CONCEPTS you should learn from this unit.

1. Definition of a determinant as a number associated with a square matrix.

2. The determinant of a 2×2 matrix is

$$\begin{vmatrix} a_{11} & a_{12} \\ a_{21} & a_{22} \end{vmatrix} = a_{11}a_{22} - a_{21}a_{12}$$

3. The minor of an element a_{ij} in a matrix is the matrix remaining after the row and column in which a_{ij} is located (that is, row i and column j) are deleted.

4. The cofactor of an element a_{ij} is the product of the minor of a_{ij} times $(-1)^{i+j}$.

5. A determinant of order higher than 2×2 is the sum of all the products found by multiplying each element in a row or a column by its cofactor. This process is often called evaluating a determinant or expanding a determinant.

6. Cramer's rule is a method of solving systems of linear equations. Here it is applied to a system of two equations, but the rule extends to three or more equations:

$$x = \frac{|D_1|}{|D|}, \quad y = \frac{|D_2|}{|D|}$$

with $|D|$ defined as the determinant of the coefficients; $|D_1|$ is $|D|$ with its first column replaced with the column of the right-hand sides of the equations (that is, the column of constant terms); $|D_2|$ is $|D|$ with its second column replaced with the column of right-hand sides of the equations.

WHAT YOU SHOULD KNOW HOW TO DO upon completion of this unit.

1. Define determinant, minor, cofactor.

2. Find determinants of the second, third, and fourth order (that is, evaluate these kinds of determinants).

3. Solve systems of linear equations by means of determinants.

SOLUTIONS to Exercise Set 7.4

2. $\begin{vmatrix} 3 & 4 \\ -1 & 2 \end{vmatrix} = (3)\cdot(2) - (-1)\cdot(4) = 6 + 4 = 10$

5. $0\cdot 3 = 1\cdot 0 = 0$

8. Computing the minors:

(a) a_{12} : $4\cdot 0 + 5\cdot 3 = 15$

(b) a_{22} : $3\cdot 0 - 5\cdot 2 = -10$

(c) a_{23} : $3\cdot(-2) - (5)\cdot(-1) = -1$

(d) a_{32} : $3\cdot(-3) - 4\cdot 2 = -17$

11. $4\cdot 8 - 5\cdot(-8) + 2\cdot(-10) = 52$

14. $-1\cdot(21 - 7) + 2\cdot(21 - 14) = 0$

using the first column.

17. Using the third row we have

$$2\begin{vmatrix} 3 & 2 & 0 \\ -1 & -3 & 0 \\ 4 & 1 & 3 \end{vmatrix} - 2\begin{vmatrix} 3 & 2 & 1 \\ -1 & -3 & -1 \\ 4 & 1 & 3 \end{vmatrix} = 2(-27 + 6) - 2(-24 + 5 + 4) = 2(-21) - 2(-15) = -12$$

20. Using the first column we have

$$1\begin{vmatrix} -1 & 4 & -1 \\ 3 & 1 & -4 \\ 2 & 0 & 2 \end{vmatrix} - 2\begin{vmatrix} 1 & 0 & 1 \\ -1 & 4 & -1 \\ 2 & 0 & 2 \end{vmatrix} = 1(-2 - 24 - 30) - 2(8 - 8) = -56$$

23.

$$x = \frac{|D_1|}{|D|} = \frac{\begin{vmatrix} 9 & 1 & -1 \\ -3 & -2 & 2 \\ 11 & 3 & 4 \end{vmatrix}}{\begin{vmatrix} 2 & 1 & -1 \\ 1 & -2 & 2 \\ 3 & 3 & 4 \end{vmatrix}} = \frac{-105}{-35} = 3$$

$$y = \frac{|D_2|}{|D|} = \frac{\begin{vmatrix} 2 & 9 & -1 \\ 1 & -3 & 2 \\ 3 & 11 & 4 \end{vmatrix}}{-35} = \frac{-70}{-35} = 2$$

23. continued

$$z = \frac{|D_3|}{|D|} = \frac{\begin{vmatrix} 2 & 1 & 9 \\ 1 & -2 & -3 \\ 3 & 3 & 11 \end{vmatrix}}{-35} = \frac{35}{-35} = -1$$

26.

$$x = \frac{|D_1|}{|D|} = \frac{\begin{vmatrix} -1 & 1 & -1 \\ 3 & -1 & 2 \\ 0 & 2 & -1 \end{vmatrix}}{\begin{vmatrix} 4 & 1 & -1 \\ 1 & -1 & 2 \\ -1 & 2 & -1 \end{vmatrix}} = \frac{-1(1 - 4) - 3(-1 + 2)}{4(1 - 4) - 1(-1 + 2) - 1(2 - 1)} = \frac{0}{-14} = 0$$

$$y = \frac{|D_2|}{|D|} = \frac{\begin{vmatrix} 4 & -1 & -1 \\ 1 & 3 & 2 \\ -1 & 0 & -1 \end{vmatrix}}{|D|} = \frac{4(-3) - 1(1) - 1(-2 + 3)}{-14} = \frac{-14}{-14} = 1$$

$$z = \frac{|D_3|}{|D|} = \frac{\begin{vmatrix} 4 & 1 & -1 \\ 1 & -1 & 3 \\ -1 & 2 & 0 \end{vmatrix}}{D} = \frac{4(-6) - 1(2) - 1(3 - 1)}{-14} = \frac{-28}{-14} = 2$$

29. $$\begin{vmatrix} a_1 + b_1 & a_2 + b_2 \\ c & d \end{vmatrix} = d(a_1 + b_1) - c(a_2 + b_2) = a_1d + b_1d - a_2c - b_2c$$

$$\begin{vmatrix} a_1 & a_2 \\ c & d \end{vmatrix} = a_1d - a_2c\ ; \qquad \begin{vmatrix} b_1 & b_2 \\ c & d \end{vmatrix} = b_1d = b_2c$$

Then adding the last two results we obtain $a_1d + b_1d - a_2c - b_2c$, the same as the first result.

32. If

$$A = \begin{bmatrix} a_{11} & a_{12} \cdots & a_{1n} \\ \vdots & & \vdots \\ a_{n1} & \cdots & a_{nn} \end{bmatrix}, \quad B = \begin{bmatrix} ka_{11} & ka_{12} \cdots & ka_{1n} \\ a_{21} & a_{22} \cdots & a_{2n} \\ \vdots & & \vdots \\ a_{n1} & \cdots & a_{nn} \end{bmatrix},$$

32. continued

then

$$|B| = Ka_{11}\begin{vmatrix}\text{cofactor}\\ \text{of } Ka_{11}\end{vmatrix} - Ka_{12}\begin{vmatrix}\text{cofactor}\\ \text{of } Ka_{12}\end{vmatrix} + \dots Ka_{1n}\begin{vmatrix}\text{cofactor}\\ \text{of } Ka_{1n}\end{vmatrix}$$

$$|B| = K\left[a_{11}\begin{vmatrix}\text{cofactor}\\ \text{of } Ka_{11}\end{vmatrix} - a_{12}\begin{vmatrix}\text{cofactor}\\ \text{of } Ka_{12}\end{vmatrix} + \dots a_{1n}\begin{vmatrix}\text{cofactor}\\ \text{of } Ka_{1n}\end{vmatrix}\right]$$

But this is precisely $K|A|$, since the cofactor of Ka_{ij} = cofactor of a_{ij} and we have proved that $|B| = K|A|$.

SOLUTIONS to Review Exercises Chapter 7

2. For matrix A, $a_{24} = -1$.

5. $\begin{bmatrix} 3 & -7 \\ 1 & 4 \end{bmatrix}$

↑ (second column) column of y coefficients

↑ (first column) column of x coefficients

8. $$\begin{bmatrix} -2 & 4 & 5 & 0 \\ 6 & -9 & 4 & 0 \\ 3 & 2 & -1 & 0 \end{bmatrix}$$

implies the system

$$\begin{aligned} -2x + 4y + 5z &= 0 \\ 6x - 9y + 4z &= 0 \\ 3x + 2y - z &= 0 \end{aligned}$$

11. The matrix $\begin{bmatrix} 1 & -4 & 2 & -18 \\ 0 & 1 & -2 & 5 \\ 0 & 0 & 1 & -1 \end{bmatrix}$

shows that $z = -1$; then the second row is

$y - 2(-1) = 5$ or $y = 3$

and the first row is

$x - 4(3) + 2(-1) = -18$ or $x = -4$

14. The augmented matrix of the system is

$$\begin{bmatrix} 3 & -1 & -17 \\ 2 & 3 & -4 \end{bmatrix}, \begin{bmatrix} 1 & -4 & -13 \\ 2 & 3 & -4 \end{bmatrix}, \begin{bmatrix} 1 & -4 & -13 \\ 0 & 11 & 22 \end{bmatrix}, \begin{bmatrix} 1 & -4 & -13 \\ 0 & 1 & 2 \end{bmatrix},$$

so that $y = 2$. Back substituting yields $x - 4(2) = -13$, $x = -5$

17. By the definition of matrix equality $2x = -6$, and so $x = -3$.

20. $$B - A = \begin{bmatrix} 2 & -1 \\ 3 & 2 \end{bmatrix} - \begin{bmatrix} -1 & 5 \\ 4 & -3 \end{bmatrix} = \begin{bmatrix} 3 & -6 \\ -1 & 5 \end{bmatrix}$$

23. $$CD = \begin{bmatrix} -1 & 0 \\ 0 & 4 \\ 2 & -2 \end{bmatrix} \cdot \begin{bmatrix} 1 & 3 & 4 \\ -1 & 0 & -6 \end{bmatrix} = \begin{bmatrix} -1 & -3 & -4 \\ -4 & 0 & -24 \\ 4 & 6 & 20 \end{bmatrix}$$

26. $$CB = \begin{bmatrix} -1 & 0 \\ 0 & 4 \\ 2 & -2 \end{bmatrix} \cdot \begin{bmatrix} -1 & 5 \\ 4 & -3 \end{bmatrix} = \begin{bmatrix} 1+0 & -5+0 \\ 0+16 & 0-12 \\ -2-8 & 10+6 \end{bmatrix} = \begin{bmatrix} 1 & -5 \\ 16 & -12 \\ -10 & 16 \end{bmatrix}$$

29. $$\begin{bmatrix} -2 & 3 & 1 & 0 \\ 1 & 4 & 0 & 1 \end{bmatrix}, \begin{bmatrix} 1 & 4 & 0 & 1 \\ -2 & 3 & 1 & 0 \end{bmatrix}, \begin{bmatrix} 1 & 4 & 0 & 1 \\ 0 & 11 & 1 & 2 \end{bmatrix}, \begin{bmatrix} 1 & 4 & 0 & 1 \\ 0 & 1 & 1/11 & 2/11 \end{bmatrix},$$

$$\begin{bmatrix} 1 & 0 & -4/11 & 3/11 \\ 0 & 1 & 1/11 & 2/11 \end{bmatrix}$$

with the right 2×2 being the inverse required.

32. $$\begin{bmatrix} 1 & 2 & -2 & 1 & 0 & 0 \\ 3 & -1 & 0 & 0 & 1 & 0 \\ 0 & 1 & 4 & 0 & 0 & 1 \end{bmatrix}, \begin{bmatrix} 1 & 2 & -2 & 1 & 0 & 0 \\ 0 & -7 & 6 & -3 & 1 & 0 \\ 0 & 1 & 4 & 0 & 0 & 1 \end{bmatrix}, \begin{bmatrix} 1 & 2 & -2 & 1 & 0 & 0 \\ 0 & 1 & 4 & 0 & 0 & 1 \\ 0 & -7 & 6 & -3 & 1 & 0 \end{bmatrix},$$

$$\begin{bmatrix} 1 & 2 & -2 & 1 & 0 & 0 \\ 0 & 1 & 4 & 0 & 0 & 1 \\ 0 & 0 & 34 & -3 & 1 & 7 \end{bmatrix}, \begin{bmatrix} 1 & 2 & -2 & 1 & 0 & 0 \\ 0 & 1 & 4 & 0 & 0 & 1 \\ 0 & 0 & 1 & -3/34 & 1/34 & 7/34 \end{bmatrix},$$

$$\begin{bmatrix} 1 & 2 & 0 & 28/34 & 2/34 & 14/34 \\ 0 & 1 & 0 & 12/34 & -4/34 & 6/34 \\ 0 & 0 & 1 & -3/34 & 1/34 & 7/34 \end{bmatrix}, \begin{bmatrix} 1 & 0 & 0 & 4/34 & 10/34 & 2/34 \\ 0 & 1 & 0 & 12/34 & -4/34 & 6/34 \\ 0 & 0 & 1 & -3/34 & 1/34 & 7/34 \end{bmatrix}$$

32. continued

The right 3 × 3 is the inverse of the coefficient matrix, and multiplying it times the column matrix $\begin{bmatrix} -4 \\ -2 \\ 1 \end{bmatrix}$ yields $x = -1$, $y = -1$, $z = 1/2$

35. $\begin{vmatrix} 2 & -1 \\ 6 & -3 \end{vmatrix} = (2)(-3) - (6)(-1) = 0$

38. $\begin{vmatrix} 1 & 2 & -1 \\ 0 & 3 & 4 \\ 0 & 0 & -1 \end{vmatrix} = 1\begin{vmatrix} 3 & 4 \\ 0 & -1 \end{vmatrix} + 0 + 0 = -3$

using the first column.

41. $x + 2y = 2$

$2x - 7y = 48$

$$x = \frac{|D_1|}{|D|} = \frac{\begin{vmatrix} 2 & 2 \\ 48 & -7 \end{vmatrix}}{\begin{vmatrix} 1 & 2 \\ 2 & -7 \end{vmatrix}} = \frac{-110}{-11} = 10$$

$$y = \frac{\begin{vmatrix} 1 & 2 \\ 2 & 48 \end{vmatrix}}{-11} = \frac{44}{-11} = -4$$

44.

$$x = \frac{\begin{vmatrix} -5 & 3 & 1 \\ -3 & 2 & 2 \\ -2 & 1 & -2 \end{vmatrix}}{\begin{vmatrix} 2 & 3 & 1 \\ 0 & 2 & 2 \\ 4 & 1 & -2 \end{vmatrix}} = \frac{-5(-6) + 3(-7) - 2(4)}{2(-6) + 4(4)} = \frac{1}{4}$$

$$y = \frac{\begin{vmatrix} 2 & -5 & 1 \\ 0 & -3 & 2 \\ 4 & -2 & -2 \end{vmatrix}}{4} = \frac{2(10) + 4(-7)}{4} = -2$$

$$z = \frac{\begin{vmatrix} 2 & 3 & -5 \\ 0 & 2 & -3 \\ 4 & 1 & -2 \end{vmatrix}}{4} = \frac{2(-1) + 4(1)}{4} = \frac{1}{2}$$

Chapter 7 - A SHORT QUIZ

1. A matrix is a ___?___ .

2. If a matrix has dimension $m \times n$, this means it has ___?___ columns.

3. Write the 3×3 identity matrix.

4. If N is a square matrix, what is $N \cdot N^{-1}$?

5. Two matrices are equal only if ___?___ .

6. In general, can a 3×4 matrix be added to any other 3×4 matrix?

7. If M and N are matrices of the same dimension, can they be multiplied if they are not square?

8. If P and Q are square matrices, does $P \cdot Q = Q \cdot P$?

9. The augmented matrix of a system of linear equations contains the coefficient matrix plus the ___?___ as the last column.

10. In transforming $\begin{bmatrix} 2 & 1 \\ -3 & 7 \end{bmatrix}$ into $\begin{bmatrix} 2 & 1 \\ 1 & 9 \end{bmatrix}$ what operation was done?

11. The product AB of two matrices will have as many rows as ___?___ and as many columns as ___?___ .

12. If A is a 2×3 matrix and B is a 3×5 matrix, is AB defined? Is BA defined?

13. A determinant is a ___?___ associated with a ___?___ matrix.

14. If the two rows of matrix $\begin{bmatrix} 2 & 3 \\ -1 & 5 \end{bmatrix}$ are interchanged, what change occurs to the determinant?

15. In the system $\begin{matrix} ax + by = c \\ dx + ey = f \end{matrix}$, show a solution for x in terms of determinants.

16. To multiply a matrix by a scalar, we must multiply ___?___ of the matrix by the scalar.

17. What is a scalar?

18. Into what sort of matrix does the Gauss-Jordan method transform the coefficient matrix?

19. In question 18 where does the solution to the system appear?

20. To find the inverse of matrix A (order n) we first form the matrix $\left[A \mid I_n \right]$ and then by ___?___ on this entire matrix transform A into ___?___ . A^{-1} then appears on the right half.

Quiz Answers Chapter 7

1. rectangular array of numbers.

2. n

3. $\begin{bmatrix} 1 & 0 & 0 \\ 0 & 1 & 0 \\ 0 & 0 & 1 \end{bmatrix}$

4. N

5. corresponding elements are equal.

6. yes

7. no

8. no

9. constant terms

10. The second row was replaced by the sum of itself and 2 times the first row.

11. A, B

12. AB is defined; BA is not.

13. number, square

14. The sign of the determinant is changed.

15. $$x = \frac{\begin{vmatrix} c & b \\ f & e \end{vmatrix}}{\begin{vmatrix} a & b \\ d & e \end{vmatrix}}$$

16. each element

17. simply a number

18. Identity matrix of the same order

19. in the right column

20. row operations, I_n

CHAPTER 8

Roots of Polynomials

Section 8.1 Polynomial Division and Synthetic Division

BASIC CONCEPTS you should learn from this unit.

1. Dividing a polynomial by another polynomial is similar to long division in arithmetic in that there is a quotient which may or may not contain a remainder.

2. Symbolically,

$$\frac{P(x)}{D(x)} = Q(x) + \frac{R(x)}{Dx}$$

   or more neatly, $P(x) = Q(x) \cdot D(x) + R(x)$ (the degree of $R(x)$ < the degree of $P(x)$).

3. Synthetic division is the name given to a "shortcut" process of long division of polynomials.

WHAT YOU SHOULD KNOW HOW TO DO upon completion of this unit.

1. Divide one polynomial by another and write the answer in the form

$$\frac{P(x)}{Dx} = Q(x) + \frac{R(x)}{D(x)}$$

2. Divide a polynomial by a polynomial of the form $(x - r)$ using synthetic division.

SOLUTIONS to Exercise Set 8.1

2.

$$\begin{array}{r} x + 1 \\ x + 2\,\overline{)\,x^2 + 3x + 3} \\ \underline{x^2 + 2x} \\ x + 3 \\ \underline{x + 2} \\ 1 \end{array}$$

Solution:

$$x + 1 + \frac{1}{x + 2}$$

5.

$$\begin{array}{r} 3x^3 - 9x^2 + 25x - 75 \\ x + 3\,\overline{)\,3x^4 - 2x^2 + 1} \\ \underline{3x^4 + 9x^3} \\ -9x^3 - 2x^2 \\ \underline{-9x^3 - 27x^2} \\ +25x^2 \\ \underline{+25x^2 + 75x} \\ -75x \\ \underline{-75x - 225} \\ +226 \end{array}$$

The solution then is:

$$3x^3 - 9x^2 + 25x - 75 + \frac{226}{x + 3}$$

8.

$$\begin{array}{r l}
 & 3x \quad + 3 \\
x^2 - x \,\big)\!\! & 3x^3 \qquad\qquad - 2x - 1 \\
 & \underline{3x^3 - 3x^2} \\
 & \qquad -3x^2 \\
 & \qquad \underline{3x^2 - 3x} \\
 & \qquad\qquad +x - 1
\end{array}$$

Solution:

$$3x + 3 + \frac{x - 1}{x^2 - 1}$$

11. $-2|$

1	-1	-6	5
	-2	+6	0
1	-3	0	5

$Q(x) = x^2 - 3x + \frac{5}{x + 2}$

or $Q(x) = x^2 - 3x + R(x) = \frac{5}{x + 2}$

14. $-3|$

1	+0	+0	+0	-81
	-3	+9	-27	+81
1	-3	+9	-27	+0

$Q(x) = x^3 - 3x^2 + 9x - 27$

$R(x) = 0$

17. $-2|$

1	+0	+0	+0	+0	+32
	-2	+4	-8	+16	-32
1	-2	+4	-8	+16	+0

$Q(x) = x^4 - 2x^3 + 4x^2 - 8x + 16$

$R(x) = 0$

20. $-3|$

8	+4	-1	-5
	-24	+60	-177
8	-20	+59	-182

$Q(x) = 8x^2 - 20x + 59$

$R(x) = \frac{-182}{x + 3}$

Section 8.2 The Remainder and Factor Theorems

BASIC CONCEPTS you should learn from this unit.

1. Remainder Theorem: If a polynomial $P(x)$ is divided by $(x - r)$, then the remainder is $P(r)$.

2. The Remainder Theorem coupled with synthetic division is an easy way to obtain (x, y) pairs for a polynomial.

3. Factor Theorem: A polynomial $P(x)$ has a factor $(x - r)$ if and only if $P(r) = 0$.

WHAT YOU SHOULD KNOW HOW TO DO upon completion of this unit.

1. For a given polynomial $P(x)$, find $P(r)$ by using the Remainder Theorem.

2. Determine by the Factor Theorem (that is, by synthetic division) whether $x - r$ is a factor of a polynomial $P(x)$.

3. Use the Remainder Theorem to assist in plotting points in graphing a polynomial.

SOLUTIONS to Exercise Set 8.2

2. $P(x) = x^4 - 3x^2 - 5x$

$P(r) = 3$

-1	1	+0	-3	-5	+0
		-1	+1	+2	+3
	1	-1	-2	-3	+3

5. $P(x) = x^6 - 3x^4 + 2x^3 + 4$

$P(r) = 0$

-1	1	+0	-3	+2	+0	+0	+4
		-1	+1	+2	-4	+4	-4
	1	-1	-2	+4	-4	+4	0

8.

-2	2	+1	+0	-5
		-4	+6	-12
	2	-3	+6	-17

R = -17

11.

-2	1	+0	+0	+0	+0	-30
		-2	+4	-8	+16	-32
	1	-2	+4	-8	+16	-62

R = -62

14. Using the Remainder Theorem we obtain

r	3 + 5 + 1 + 5	-2
1	3 8 + 9 + 14	12 = P(1)
0	...	-2 = P(0)
-1	3 + 2 - 1 + 4	-6 = P(-1)
-2	3 - 1 + 3 - 1	0 = P(-2)
-3	3 - 4 + 13 - 34	66 = P(-3)

x	y
1	12
0	-2
-1	-6
-2	0
-3	66

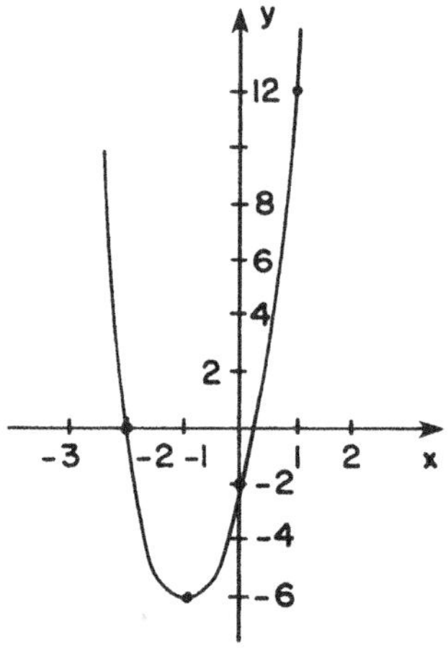

17.

r	1 - 10 + 0 + 0 + 1	
10	1 0 0 0 + 1	P(10)
6	1 - 4 - 24 - 144 - 863	P(6)
1	1 - 10 - 10 - 10 - 9	P(1)
0	... + 1	P(0)
-1	1 - 11 + 11 - 11 + 12	P(-1)
-4	1 - 6 + 24 - 96 + 396	P(-4)

x	y
10	1
6	-863
1	-9
0	+1
-1	+12
-4	+396

(continued on page 204)

17. continued

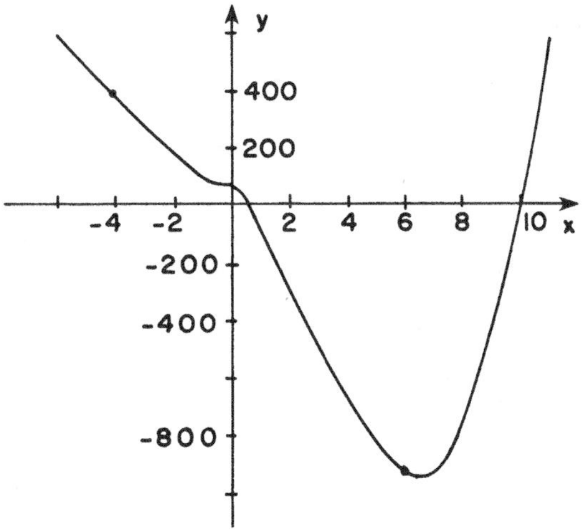

20. $$\begin{array}{r|rrrr} 1 & 1 & +4 & -3 & +1 \\ & & 1 & +5 & +2 \\ \hline & 1 & +5 & +2 & +3 \end{array}$$

$R = 3$, so $x - 1$ is <u>not</u> a factor.

23. $$\begin{array}{r|rrrr} -3 & 1 & +0 & +0 & +27 \\ & & -3 & +9 & -27 \\ \hline & 1 & -3 & +9 & 0 \end{array}$$

$R = 0$, so $x + 3$ is a factor.

26. $$\begin{array}{r|rrrr} 3 & 1 & +0 & +0 & +27 \\ & & +3 & +9 & +27 \\ \hline & 1 & +3 & +9 & +54 \end{array}$$

$R = 54$, so $x - 3$ is <u>not</u> a factor

29. We synthetically divide and determine K so that the remainder $= 0$.

$$\begin{array}{r|rrrr} 2 & 1 & -3 & +K & -1 \\ & & +2 & -2 & 2(K-2) \\ \hline & 1 & +1 & +(K-2) & \end{array}$$

$$2(K-2) = 1$$
$$2K - 4 = 1$$
$$2K = 5$$
$$\underline{K = 5/2}$$

32. $$\begin{array}{r|rrrrr} r & 2 + & 0 + 3 & + & 0 & 2 \\ & & 2r + 2r^2 & & 2r^3+3r & 2r^4+3r^2 \\ \hline & 2 & +2r + 2r^3+3 & + & 2r^3+3r & 0 \end{array}$$

For the sum of $2 + (2r^4 + 3r^2)$ to equal 0 we have the equation

$$2r^4 + 3r^2 = -2$$

which is impossible, since $r^4 > 0$ and $r^2 > 0$ for all real r.

BASIC CONCEPTS you should learn from this unit.

1. The <u>conjugate</u> of a complex number $a + bi$ is $a - bi$.

2. The product of two conjugate complex numbers is a real number; for example, $(a + bi)(a - bi) = a^2 + b^2$.

3. The quotient of two complex numbers is found by multiplying the dividend and the divisor by the conjugate of the divisor.

4. The reciprocal of a complex number $(a + bi)$ is found by division:

$$\frac{1}{a+bi}\cdot\frac{(a-bi)}{(a-bi)} = \frac{a-bi}{a^2+b^2} \quad \text{or} \quad \frac{a}{a^2+b^2} - \frac{bi}{a^2+b^2}$$

5. Five basic properties of complex numbers are listed here, using the symbols Z and $\overline{Z}$ to represent a complex number and its conjugate, respectively.

   A. $\overline{Z} = \overline{W}$ if and only if $Z = W$.

   B. $\overline{Z} = Z$ if and only if Z is real.

   C. $\overline{Z + W} = \overline{Z} + \overline{W}$.

   D. $\overline{Z \cdot W} = \overline{Z} \cdot \overline{W}$.

   E. $\overline{Z^n} = \overline{Z}^n$, n a positive integer.

6. Factor Theorem: r is a root of $P(x)$ if and only if $x - r$ is a factor of $P(x)$.

7. Fundamental Theorem of Algebra: Every polynomial $P(x)$ of degree ≥ 1 has at least one complex root.

8. Linear Factor Theorem: A polynomial $P(x)$ of degree $n \geq 1$ can be written as the product of n linear factors such that $P(x) = a(x - r_1)(x - r_2) \ldots (x - r_n)$.

9. Number of Roots Theorem: If $P(x)$ is a polynomial of degree $n \geq 1$, then $P(x)$ has exactly n roots among the complex numbers.

10. The same number may occur as a root of a polynomial more than once, say K times. It then is said to have multiplicity K. For example, in

$$P(x) = x^3 + 3x^2 + 3x + 1 = (x + 1)^3$$

    the number -1 is a root of multiplicity 3.

11. The "depressed" equation $Q(x) = 0$, taken from

$$P(x) = (x - r)Q(x)$$

    can be used to advantage in finding the roots of $P(x)$.

12. Conjugate Root Theorem: If $P(x)$ is a polynomial of degree $n \geq 1$ with real coefficients and if $a + bi$ $(b \neq 0)$ is a root of $P(x)$, then $a - bi$ is also a root. (In common language: Complex roots occur in conjugate pairs.)

WHAT YOU SHOULD KNOW HOW TO DO upon completion of this unit.

1. Multiply a complex number by its conjugate.

2. Divide two complex numbers.

3. Find the reciprocal of a complex number.

4. Write a polynomial $P(x)$ if given its roots.

5. Find the roots of a polynomial if given some of the roots by using the Factor Theorem, Conjugate Root Theorem, and Number of Roots Theorem.

SOLUTIONS to Exercise Set 8.3

2. $(3 + i)(3 - i) = 9 - i^2 = 9 + 1 = 10$

5. $(-4 - 2i)(-4 + 2i) = 16 - 4i^2 = 16 + 4 = 20$

8. $\frac{(1 + 3i)}{(2 - 5i)} \cdot \frac{(2 + 5i)}{(2 + 5i)} = \frac{2 + 11i - 13i^2}{4 + 25} = \frac{-13 + 11i}{29} = \frac{-13}{29} + \frac{11i}{29}$

11. $\frac{(3 - 2i)}{(2 - i)} \cdot \frac{(2 + i)}{(2 + i)} = \frac{6 - i - 2i^2}{4 + 1} = \frac{8 - i}{5} = \frac{8}{5} - \frac{i}{5}$

14. $\frac{(5 - 2i)}{-3i} \cdot \frac{+3i}{+3i} = \frac{15i - 6i^2}{-9i^2} = \frac{+6 + 15i}{9} = \frac{2}{3} + \frac{5i}{3}$

17. $\frac{1}{(4 + 3i)} \cdot \frac{4 - 3i}{(4 - 3i)} = \frac{4 - 3i}{16 + 9} = \frac{4}{25} - \frac{3i}{25}$

20. $\frac{1}{-7i} \cdot \frac{7i}{7i} = \frac{7i}{49} = \frac{i}{7}$

23. Let $Z = a + bi$, then $\overline{Z} = a - bi$;
Let $W = c + di$, then $\overline{W} = c - di$.

$Z \cdot W = (a + bi)(c + di) = ac - bd + (ad + bc)i$

and so $\overline{Z \cdot W} = ac - bd - (ad + bc)i$

Also $\overline{Z} \cdot \overline{W} = (a - bi)(c - di) = ac - bd - adi - bci = (ac - bd) - (ad + bc)i$

and we have proved that $\overline{Z \cdot W} = \overline{Z} \cdot \overline{W}$.

26. $P(x)$ must equal

$$(x - 5)(x + 5)(x - 1)(x + 1) = (x^2 - 25)(x^2 - 1)$$

or $P(x) = x^4 - 26x^2 + 25$ (using the Factor Theorem and the Linear Factor Theorem).

29. $P(x) = (x - 4)[x - (1 - \sqrt{3})]\,[x - (1 + \sqrt{3})] = (x - 4)(x^2 - 2x - 2)$

$= x^3 - 6x^2 + 6x + 8$

32. $P(x) = (x - 3)^2(x + 2)(x - 2) = (x^2 - 6x + 9)(x^2 - 4)$

$= x^4 - 4x^2 - 6x^3 + 24x + 9x^2 - 36 = x^4 - 6x^3 + 5x^2 + 24x - 36$

Using synthetic division to find $P(4)$, we obtain

$$\begin{array}{r|rrrrr} 4 & 1 & -6 & +5 & +24 & -36 \\ & & 4 & -8 & -12 & +48 \\ \hline & 1 & -2 & -3 & +12 & +12 \end{array}$$

Since $P(4) = 12$ as required, the number $a = 1$ and $P(x) = x^4 - 6x^3 + 5x^2 + 24x - 36$.

35. Given $(x - 3)(x + 1)(x - 2) = 0$ the roots can be read directly as $3, -1, 2$.

38. $(x^2 - x)(x^2 - 2x + 5) = 0$. From $x^2 - x = 0$, we get $x(x - 1) = 0$ so $x = 0$, $x = 1$ and from $x^2 - 2x + 5 = 0$ we get $x = \frac{2 \pm \sqrt{4 - 20}}{2} = 1 \pm 2i$. So the roots of the equation are $0, 1, 1+2i, 1-2i$.

41. $(x - 5)^3(x + 5)^2 = 0$ directly tells us that 5 is a root of multiplicity 3 and -5 is a root of multiplicity 2 .

44. $(x - 1)^2(x + 4)$ will do it; or

$(x^2 - 2x + 1)(x + 4) = x^3 + 4x^2 - 2x^2 - 8x + x + 4$

$= x^3 + 2x^2 - 7x + 4$

47. Using synthetic division is one way.

$$\begin{array}{r|rrrr} -1 & 1 & +0 & -3 & -2 \\ & & -1 & +1 & +2 \\ \hline & 1 & -1 & -2 & 0 \end{array}$$

So the depressed equation is

$x^2 - x - 2 = 0$

which factors to $(x - 2)(x + 1) = 0$, so that $x = 2$, $x = -1$ are the other two roots.

50. By synthetic division

$$\begin{array}{r|rrrr} -1 & 1 & -2 & -7 & -4 \\ & & -1 & +3 & +4 \\ \hline & 1 & -3 & -4 & 0 \end{array}$$

we get the depressed equation $x^2 - 3x - 4 = 0$. The left-hand side factors:

$(x - 4)(x + 1) = 0$

so the other roots are $x = 4$, $x = -1$.

53. If $1 + 3i$, -2 are the only roots, then $[x - (1 + 3i)][x - (-2)] = 0$ is the polynomial equation, so

$$P(x) = x^2 - (1 + 3i)x + 2x - 2 - 6i$$

or

$$P(x) = x^2 + (1 - 3i)x - (2 + 6i)$$

56. As in solution 53 we can write

$$P(x) = (x + 2)(x - 3)(x - 1 - 2i) = x^3 - (2 + 2i)x^2 - (5 - 2i)x + 6 + 12i$$

59. If $(3 - i)$ is a root, then so is $(3 + i)$. Multiplying $[x - (3 - i)] \cdot [x - (3 + i)] = x^2 - 6x + 10$, which must divide the given polynomial. Doing this we obtain

$$\begin{array}{r|l} & x - 1 \\ \hline x^2 - 6x + 10 & x^3 - 7x^2 + 16x - 10 \\ & \underline{x^3 - 6x^2 + 10x} \\ & \quad -x + 6x - 10 \\ & \quad \underline{-x + 6x - 10} \\ & \qquad\qquad 0 \end{array}$$

Now, the required product is $(x^2 - 6x + 10)(x - 1)$.

62. If $(-2 + i)$ is a root, then so is $(-2 - i)$ and multiplying $[x - (-2 + i)] \cdot [x - (-2 - i)]$ yields $x^2 + 4x + 5$. Dividing this into the given polynomial, we obtain

$$\begin{array}{r|l} & x^2 - x - 6 \\ \hline x^2 + 4x + 5 & x^4 + 3x^3 - 5x^2 - 29x - 30 \\ & \underline{x^4 + 4x^3 + 5x^2} \\ & \quad -x^3 - 10x^2 - 29x \\ & \quad \underline{-x^3 - 4x^2 - 5x} \\ & \qquad -6x^2 - 24x - 30 \\ & \qquad \underline{-6x^2 - 24x - 30} \end{array}$$

so that

$$(x^2 + 4x + 5)(x^2 - x - 6) \quad \text{or} \quad (x^2 + 4x + 5)(x - 3)(x + 2)$$

is the required product.

65. $x - a - bi = P(x)$; or $x - (a + bi) = P(x)$

BASIC CONCEPTS you should learn from this unit.

1. Descartes' Rule of Signs:

   A. The number of positive roots of $P(x)$ is either equal to the number of variations in sign of $P(x)$ or less than that number by an even number.

   B. The number of negative roots of $P(x)$ is either equal to the number of variations in sign of $P(-x)$ of less than that number by an even number.

2. The Rational Root Theorem: If the coefficients of a polynomial

   $P(x) = a_n x^n + a_{n-1} x^{n-1} + \ldots + a_0$, $a_n \neq 0$, and if p/q is a rational root in lowest terms, then p divides a_0 and q dividies a_n .

3. A rational root may be a multiple root.

WHAT YOU SHOULD KNOW HOW TO DO upon completion of this unit.

1. Describe the nature of the roots of a polynomial using Descartes' Rule of Signs.

2. Find all the rational roots of a polynomial equation using Descartes' Rule of Signs, the Rational Root Theorem, and the Conjugate Root Theorem.

SOLUTIONS to Exercise Set 8.4

2. $2x^6 + 5x^5 + x^3 - 6 = 0$; one change of sign, so one positive root.

   $P(-x) = 2x^6 - 5x^5 - x^3 - 6$; one change of sign, so one negative root.

   one positive, one negative, four complex roots

5. $x^5 - 4x^3 + 7x - 4 = 0$; three changes of sign, so there are either three positive roots or one positive root.

   $P(-x) = -x^5 + 4x^3 - 7x - 4$; two changes of sign, so there are two or no negative roots. Four possibilities exist: (1) three positive, two negative roots; (2) three positive, zero negative, and two complex roots; (3) one positive, two negative, and two complex roots; (4) one positive and four complex roots.

8. $x^5 + 6x^4 - x^3 - 2x - 3 = 0$; one change of sign, so there is exactly one positive root.

   $P(-x) = -x^5 + 6x^4 + x^3 + 2x - 3 = 0$; two changes of sign, so there are two negative roots or none. The possibilities are one positive, two negative, and two complex roots; or one positive, zero negative, and four complex roots.

11. $x^8 + 7x^3 + 3x - 5 = 0$; one change of sign, so there is one positive root.

    $P(-x) = x^8 - 7x^3 - 3x - 5 = 0$; one change of sign, so there is one negative root. The answer is one positive, one negative, and six complex roots.

14. $3x^3 - x^2 - 3x + 1 = 0$; Since $a_n = 3$ and $a_0 = 1$, the possible rational roots are ± 1, $\pm 1/3$.

	3	-1	-3	+1
$\frac{1}{3}$	3	0	-3	0

This synthetic division shows that 1/3 is a root, and the depressed equation $3x^2 - 3 = 0$ has +1 and -1 for roots. Thus the roots are 1/3 , +1 , -1 .

17. $5x^6 - x^5 - 5x^4 + 6x - x^2 - 5x + 1 = 0$. The possibilities are ± 1 and $\pm 1/5$. Using synthetic division, we obtain

	5	-1	-5	+6	-1	-5	+1	
1	5	4	-1	+5	+4	-1	0	1 is a root
-1	5	-1	0	5	-1	0		-1 is a root
1/5	5	0	0	5	0			1/5 is a root
-1	5	-5	+5	0				-1 is a root

From $5x^2 - 5x + 5 = 0$ we see that the roots of this depressed equation are complex, so we have located all the rational roots.

20. $6x^4 + 2x^3 + 7x^2 + x + 2 = 0$; there are no changes of sign, so there are <u>no</u> positive roots.

$P(-x) = 6x^4 - 2x^3 + 7x^2 - x + 2 = 0$; there are four changes of sign, so there are <u>four</u>, <u>two</u>, or <u>no</u> negative roots. Possibilities are -1 , -2 , -1/2 , -1/3 , -2/3 , -1/6 . Using synthetic division, we obtain

	6	+2	+7	+1	+2
-1	6	-4	11	-10	12
-2	6	-10	+27	-53	108
-1/2	6	-1	15/2	-11/4	27/8
-1/3	6	0	7	-4/3	22/9
-1/6	6	+1	+41/6	-5/36	437/216
-2/3	6	-2	25/3	-41/9	136/27

Since no zero remainders occurred, we conclude that there are <u>no</u> rational roots.

23. $4x^4 + x^3 + x^2 + x - 3 = 0$; one positive root, so there are three or one negative roots. Possibilities are ± 1, ± 3, $\pm 1/2$, $\pm 3/2$, $\pm 1/4$, $\pm 3/4$.

	4	+1	+1	+1	-3	
1	4	5	6	7	4	
1/2	4	3	5/2	9/4	-15/8	
3/4	4	4	4	4	0	3/4 is a root
-1	4	0	4	0		-1 is a root

23. continued

Then the depressed equation $4x^2 + 4 = 0$ gives the two remaining roots: $4x^2 = -4$, $x^2 = -1$, $x = \pm i$. The roots are $3/4$, -1, $\pm i$.

26. $12x^4 - 8x^3 - 75x^2 - 16x - 5 = 0$; there is one positive root and three or one negative roots. Possible roots are ± 1, ± 5, $\pm 1/2$, $\pm 1/3$, $\pm 1/4$, $\pm 1/6$, $\pm 1/12$, $\pm 5/2$, $\pm 5/3$, $\pm 5/4$, $\pm 5/6$, $\pm 5/12$.

Testing these by synthetic division reveals no rational roots. Thus the roots must be irrational or complex.

29. $2x^4 - x^3 - 28x^2 + 30x - 8 = 0$. Possible rational roots are ± 1, ± 2, ± 4, ± 8, $\pm 1/2$.

	2	-1	-28	+30	-8	
1/2	2	0	-28	16	0	1/2 is a root
-4	2	-8	4	0		-4 is a root

The depressed equation $2x^2 - 8x + 4 = 0$ or $x^2 - 4x + 2 = 0$ gives

$$x = \frac{4 \pm \sqrt{8}}{2} = 2 \pm \sqrt{2}$$

Roots are $\frac{1}{2}$, -4, $2 \pm \sqrt{2}$.

32.

	1	-4	-k	+6k	+9	
3	1	-1	-3-k	-9+3k	0	if $3(-9 + 3k) = -9$, $k = 2$

so if $k = 2$, the equation becomes

$$x^4 - 4x^3 - 2x^2 + 12x + 9$$

	1	-4	-2	+12	+9	
3	1	-1	-5	-3	0	3 is a root
-3	1	-7	19	-45	+144	no root
9	1	5	43	384	+	no root
-9	1	-13	115	-	+	no root
-1	1	-5	+3	+9	0	-1 is a root
1	1	-3	-5	+7	+	no root

38. Given that p is a prime, we let $x = \sqrt{p}$, or $x^2 = p$, or $x^2 - p = 0$. Now since p is prime the only possible rational roots to this equation are ± 1 and $\pm p$. Testing these quickly reveals that none are roots. Hence the two roots must be irrational and so p is irrational.

SOLUTIONS to Review Exercises Chapter 8

2.

```
       1  -3  +0  +2  -5
  -2 | 1  -5  10 -18  31      R = 31
```

$Q(x) = x^3 - 5x^2 + 10x - 18$

5.

```
       2  +4  +3  +5  -2
  -2 | 2   0   3  -1   0
```

Since $R = 0$, $x + 2$ is a factor of $P(x)$.

8. $\dfrac{2+i}{-5i} \cdot \dfrac{5i}{5i} = \dfrac{10i + 5i^2}{-10i^2} = \dfrac{-1}{2} + i$

11. $\dfrac{1}{-4i} \cdot \dfrac{4i}{4i} = \dfrac{4i}{16} = \dfrac{i}{4}$

14. $P(x) = (x - 3)(x - \sqrt{-3})(x + \sqrt{-3}) = x^3 - 3x^2 + 3x - 9$

17. $P(x) = (x - i)(x - i)(x + i)(x + i) = (x^2 - i^2)(x^2 - i^2)$
$= (x^2 + 1)(x^2 + 1) = x^4 - 2x^2 + 1$

20.

```
       1  -2  -9  +4
   4 | 1   2  -1   0
```

so 4 is indeed a root. The depressed equation $x^2 + 2x - 1 = 0$ yields

$$x = \frac{-2 \pm \sqrt{4 + 4}}{2} = -1 \pm \sqrt{2}$$

as the other two roots.

23. $x^5 - x^4 + 3x^3 - 4x^2 + x - 5$ has five changes of sign, so there are at most five <u>positive</u> roots.

$-x^5 - x^4 - 3x^3 - 4x^2 - x - 5$ has no changes of sign, so there are no negative roots.

26. For $6x^3 - 5x^2 - 33x - 18 = 0$ the possible rational roots are ± 1, ± 2, ± 3, ± 6, ± 18, $\pm 1/2$, $\pm 3/2$, $\pm 9/2$, $\pm 1/3$, $\pm 2/3$, $\pm 1/6$. Some synthetic division

```
       6  -5  -33  -18
   3 | 6  13    6    0
```

reveals that 3 is a root. Then the depressed equation $6x^2 + 13x + 6 = 0$ yields $-3/2$ and $-2/3$ as the other two roots.

29. $6x^3 + 15x^2 - x - 10 = 0$. Trying synthetic division with the possible roots will yield

$$\begin{array}{r|rrrr} & 6 & +15 & -1 & -10 \\ -1 & 6 & 9 & -10 & 0 \end{array}$$

so -1 is a root. Now use the depressed equation $6x^2 + 9x - 10 = 0$ to obtain $x = \frac{-9 \pm \sqrt{81 + 240}}{12}$ or $\frac{-9 \pm \sqrt{321}}{12}$ as the two remaining roots.

Chapter 8 - A SHORT QUIZ

1. How many roots does a polynomial of degree n have?

2. In the following, where $P(x)$, $D(x)$, $Q(x)$, and $R(x)$ are polynomials, the degree of $R(x)$ will be ___?___ than the degree of $P(x)$.

$$\frac{P(x)}{D(x)} = Q(x) + \frac{R(x)}{D(x)}$$

3. In the expression in Problem 2, if $R(x) = 0$, we know that $D(x)$ is a ___?___ of $P(x)$.

4. In the expression

$$\frac{P(x)}{D(x)} = Q(x) + \frac{R(x)}{D(x)}$$

if $D(x) = x - r$, then $R(x)/D(x)$ is the value of $P(x)$ at $x =$ ___?___

5. If $a + bi$, a complex number with $b \neq 0$, is a root of a polynomial $P(x)$, then ___?___ is also a root of $P(x)$.

6. Multiply $(a + bi)\cdot(a - bi)$.

7. Write a polynomial $P(x)$ in factored form if its roots are known to be 6, -3, 4 .

8. If $P(x) = 3x^2 - 2x + 5$, use synthetic division to find $P(-2)$.

9. If $P(x) = 4x^3 + 3x^2 + x + 2$, how many positive roots are there? How many negative roots are possible?

10. Determine the roots of $(x-2)^4(x + 5)^3 = 0$.

11. Synthetic division of $(x^3 - x^2 - 2x + 8) \div (x + 2)$ reveals

$$\begin{array}{r|rrrr} & 1 & -1 & -2 & +8 \\ -2 & 1 & -3 & 4 & 0 \end{array}$$

showing that -2 is a root of the third degree polynomial. Use the depressed equation to find the other roots.

12. Does a polynomial always have a root in the complex numbers?

13. Complex numbers are usually expressed in the form $a + bi$. What kind of numbers are a and b ? What does i represent?

14. Write the equation with real coefficients of degree two that has $-7i$ as one of its roots.

15. If $x - r$ is a factor of $P(x)$, then ___?___ is a ___?___ of the equation $P(x) = 0$.

Quiz Answers Chapter 8

1. n

2. less than

3. factor

4. r

5. $a - bi$

6. $a^2 + b^2$

7. $P(x) = (x - 6)(x + 3)(x - 4)$

8. 21

9. no positive roots; three or one negative roots

10. 2 is a root of multiplicity 4;
    -5 is a root of multiplicity 3.

11. $\frac{3}{2} \pm \frac{i\sqrt{7}}{2}$

12. yes

13. a and b are real numbers;
    i represents $\sqrt{-1}$.

14. $x^2 + 49 = 0$

15. r, solution or root

CHAPTER 9

Topics in Algebra

Section 9.1 Arithmetic Progressions

BASIC CONCEPTS you should learn from this unit.

1. An <u>infinite sequence</u> is a function whose domain is the set of natural numbers.

2. A <u>finite</u> sequence is a sequence that terminates.

3. An arithmetic sequence (sometimes called an arithmetic progression) is a sequence such that each term differs from its predecessor by a fixed number, called the common difference.

4. Symbolically, the nth term in an arithmetic sequence can be written by:

   A. a recursive formula, $a_n = a_{n-1} + d$;

   B. the formula $a_n = a_1 + (n - 1)d$, which relates a_n to a_1.

5. Adding the terms of a sequence results in a series. Thus, adding the terms of an arithmetic sequence would yield an arithmetic series. In symbols,

$$S_n = a_1 + a_2 + \ldots + a_{n-1} + a_n$$

6. Two formulas for the sum of an arithmetic series are

   A. $S_n = \frac{n}{2}(a_1 + a_n)$;

   B. $S_n = \frac{n}{2}[2a_1 + (n - 1)d]$.

WHAT YOU SHOULD KNOW HOW TO DO upon completion of this unit.

1. Define arithmetic sequence and arithmetic series.

2. Find a_n if given a_1 and d of an arithmetic sequence.

3. Find d if given a_n and a_1 of an arithmetic sequence.

4. Find a_n if given a_{n-1} and d of an arithmetic sequence.

5. Recognize an arithmetic sequence or progression.

6. Find the sum of n terms of an arithmetic series if given sufficient information.

SOLUTIONS to Exercise Set 9.1

2. Recall that the domain is the set of natural numbers; then if $a_n = 2n + 1$, the first four terms are $2\cdot1 + 1$, $2\cdot2 + 1$, $2\cdot3 + 1$, $2\cdot4 + 1$, or 3, 5, 7, 9.

5. Using $n = 1, 2, 3, 4$ successively, we have $\frac{1}{3}, \frac{4}{5}, \frac{9}{7}, \frac{16}{9}$ as the first four terms.

8. $\frac{8}{2}, \frac{11}{2}, \frac{14}{2}, \frac{17}{2}$ is an arithmetic progression with $d = \frac{3}{2}$.

11. $0, \frac{1}{4}, \frac{2}{4}, \frac{3}{4}$ is an arithmetic sequence with $d = \frac{1}{4}$.

14. 12, 8, 4, 1 is <u>not</u> an arithmetic sequence, since there is no common difference.

17. $a_1 = 3$; $a_2 = 3 + \left(-\frac{1}{2}\right)$; $a_3 = 3 + 2\cdot\left(-\frac{1}{2}\right)$, and $a_4 = 3 + 3\left(-\frac{1}{2}\right)$. Thus we have $3, \frac{5}{2}, 2, \frac{3}{2}$.

20. Here we have 6, $6 + \frac{5}{2}$, $6 + 2\cdot\left(\frac{5}{2}\right)$, $6 + 3\left(\frac{5}{2}\right)$, or $6, \frac{17}{2}, 11, \frac{27}{2}$.

23. $a_{12} = a_1 + (12 - 1)d$, so $a_{12} = 14 + 11(-2) = -8$.

26. If $a_1 = \frac{1}{2}$ and $a_{12} = 6$, we can write $6 = \frac{1}{2} + 11\cdot d$ and find $d = \frac{1}{2}$; then $a_{30} = \frac{1}{2} + 29\cdot\left(\frac{1}{2}\right) = \frac{30}{2}$ or 15.

29. If $a_1 = -\frac{1}{4}$ and $a_{41} = 10$, we can write $10 = -\frac{1}{4} + 40d$ and find $d = \frac{41}{160}$; then $a_{22} = -\frac{1}{4} + 21\cdot\left(\frac{41}{160}\right) = \frac{821}{160}$.

32. $S_n = \frac{n}{2}[2a_1 + (n - 1)d]$, so in our case $S_{24} = \frac{24}{2}\left[2(-4) + 23\cdot\frac{1}{2}\right] = 42$.

35. Here $S_{40} = \frac{40}{2}[2(82) + 39\cdot(-2)] = 1720$.

38. In this instance $a_1 = 44$; $d = -3$, $S_n = 340$, and n is unknown. Thus we can write

$340 = \frac{n}{2}[2\cdot 44 + (n - 1)(-3)]$ and solve for n: $340 = \frac{n}{2}(91 - 3n)$; $680 = 91n - 3n^2$;

and then $3n^2 - 91n + 680 = 0$ and $(3n - 40)(n - 17) = 0$ and thus $n = 17$.

41. $a_1 = \frac{1}{2}$, $a_n = \frac{7}{4}$, $S_n = \frac{27}{4}$. We use the formulas to make two equations:

(1) $\frac{27}{4} = \frac{n}{2}[1 + (n - 1)d]$

(2) $\frac{7}{4} = \frac{1}{2} + (n - 1)d$

We solve for d in (2):

$$d = \frac{5}{4(n - 1)}$$

and substitute this result in (1):

$$\frac{27}{4} = \frac{n}{2} + \frac{n}{2}\cdot \cancel{(n-1)}\,\frac{5}{4\cancel{(n-1)}}$$

$$\frac{27}{4} = \frac{n}{2} + \frac{5n}{8}$$

$$54 = 4n + 5n$$

$$n = 6$$

44. Since $a_3 = 9$ and $a_6 = 18$, the common difference $d = (18 - 9)/3 = 3$. Subtracting 2d or 6 from a_3, that is, $9 - 6 = 3$, we get $a_1 = 3$. Now $S_{12} = \frac{12}{2}[2 \cdot 3 + (11)(3)] = 234$.

Section 9.2 Geometric Progressions

BASIC CONCEPTS you should learn from this unit.

1. A geometric sequence or progression is a sequence such that each term is obtained by multiplying the previous term by a fixed number. In symbols, $a_n = r \cdot a_{n-1}$ for $n > 1$; r is called the common ratio.

2. The nth term of a geometric sequence or progression is given by $a_n = a_1 r^{n-1}$.

3. Geometric means are the terms between the first and last terms.

4. A geometric series is the sum of the terms in a geometric progression.

5. $S_n = \frac{a_1(1 - r^n)}{1 - r}$ is the formula for the sum of a geometric series.

6. If $|r| < 1$, then r^n gets closer and closer to zero as n gets larger, and thus the sum of an infinite geometric series with $|r| < 1$ becomes

$$S_n = \frac{a_1}{1 - r}$$

WHAT YOU SHOULD KNOW HOW TO DO upon completion of this unit.

1. Define geometric sequence and geometric series.

2. Find a_n if given a_1 and r of a geometric sequence.

3. Find r if given several terms of a geometric sequence.

4. Insert geometric means between two terms of a geometric sequence.

5. Recognize a geometric sequence.

6. Find the sum of n terms of a geometric series.

7. Find the sum of an infinite geometric series that has $|r| < 1$.

SOLUTIONS to Exercise Set 9.2

2. $(-3) \cdot (-4) = 12$; $(-3) \cdot 12 = -36$; $(-3)(-36) = 108$; so yes, this is a geometric sequence; $r = -3$.

5. Yes, it is geometric with $r = 0.2$.

8. $a_1 = -4$, $a_2 = (-4)\cdot 2 = -8$, and so on, yielding $-4, -8, -16, -32, \ldots$.

11. $-3, -6, -12, -24$ are found by multiplying a term's predecessor by 2, the common ratio.

14. $a_n = a_1 \cdot r^{n-1}$, so $a_6 = 18 \cdot \left(-\frac{1}{2}\right)^5 = 18\left(-\frac{1}{32}\right) = -\frac{9}{16}$.

17. $a_5 = a_1 \cdot r^4$, so $\frac{1}{27} = 3 \cdot r^4$ and then $r^4 = \frac{1}{81}$ and $r = \frac{1}{3}$ or $-\frac{1}{3}$. Then

$a_7 = 3 \cdot \left(\frac{1}{3}\right)^6 = \frac{1}{243}$.

20. $a_7 = a_4 \cdot r^3$, so $1 = \frac{1}{4}\cdot r^3$, $r^3 = 4$; $r = \sqrt[3]{4}$.

23. $a_n = a_1 r^{n-1}$, so $32 = \frac{1}{2} r^{n-1}$, and $64 = r^{n-1}$. But $r = 2$, and thus $2^6 = 2^{n-1}$ or $n = 7$.

26. We would write -3, a_2, a_3, 192, and find r from $192 = -3r^3$, $r^3 = -64$, and finally $r = -4$. Using this value to find a_2, we get $a_2 = -3\cdot(-4) = 12$, and then $12 \cdot (-4) = -48$ for a_3.

29. $$S_n = \frac{a_1(1 - r^n)}{1 - r}$$

so with $a_1 = 3$ and $r = \frac{1}{3}$ we have

$$S_7 = \frac{3[1 - (1/3)^7]}{1 - 1/3} = \frac{9[1 - (1/3^7)]}{2} = \frac{9 - (1/3^5)}{2} = \frac{2186}{486} = \frac{1093}{243}\ .$$

32. $$S_6 = \frac{2\left(1 - \frac{2}{3}\right)^6}{1 - \frac{2}{3}} = 6\left(1 - \frac{64}{729}\right) = 6\left(\frac{665}{729}\right) = \frac{1330}{243}$$

35. $a_4 = a_1 \cdot r^3$, so $-\frac{54}{8} = 2r^3$, $r^3 = -\frac{27}{8}$, and $r = -\frac{3}{2}$. Now

$$S_5 = \frac{2[1 - (-3/2)^5]}{1 - (-3/2)} = \frac{4}{5}\left[\left(1 + \frac{243}{32}\right)\right] = \frac{4}{5}\left(\frac{275}{32}\right) = \frac{25}{8}$$

38. Here $a_1 = 20000$, $r = 1.05$, and we seek a_{11}.

$a_{10} = 20000(1.05)^{10} = 32577.9 = 32578$ people

41. $a_1 = 1$, $r = \frac{1}{2}$, $S = \dfrac{1}{1 - \frac{1}{2}} = 2$.

44. $a_1 = \frac{1}{2}$, $r = -\frac{1}{2}$, $S = \frac{0.5}{1 + .5} = \frac{0.5}{1.5} = \frac{1}{3}$.

47. $a_1 = 0.5$, $r = 0.5$, $s = \frac{0.5}{1 - 0.5} = 1$.

50. $3.6\overline{6} = 3 + 0.6 + 0.06 + 0.006 + \ldots$.

If we choose $0.6 = a_1$ and $r = 0.1$, the sum from 0.6 on is $\frac{0.6}{1 - 0.1} = \frac{2}{3}$ and the final sum is $3\frac{2}{3}$ or $\frac{11}{3}$.

53. Writing $0.325\overline{325}$ as $0.325 + 0.000325 + \ldots$, we have $a_1 = 0.325$ and $r = 0.001$. Then the sum becomes $\frac{0.325}{1 - 0.001} = \frac{325}{999}$.

Section 9.3 Mathematical Induction

BASIC CONCEPTS you should learn from this unit.

1. The principle of mathematical induction is a method of proving certain statements involving the natural numbers.

2. Mathematical induction is defined as follows: If a statement involving the natural numbers, n,

   A. is true when $n = 1$, and

   B. whenever the statement is true for some natural number k, it is also true for the natural number $k + 1$, then

   the statement is true for all n.

WHAT YOU SHOULD KNOW HOW TO DO upon completion of this unit.

1. Define mathematical induction.

2. Prove certain types of statements by mathematical induction.

SOLUTIONS to Exercise Set 9.3

2. To prove $1^2 + 3^2 + 5^2 \ldots + (2n - 1)^2 = \frac{n(2n + 1)(2n - 1)}{3}$

   Step 1. Evaluating the formula when $n = 1$ yields

$$\frac{1(2 \cdot 1 + 1)(2 \cdot 1 - 1)}{3} = 1$$

   which does equal 1^2, which verifies the formula when $n = 1$.

   Step 2. For $n = k$ we have

$$1^2 + 3^2 + \ldots + (2k - 1)^2 = \frac{k(2k + 1)(2k - 1)}{3}$$

   Adding the square of the next consecutive odd integer to both sides gives

(continued on page 220)

2. continued

$$1^2 + 3^2 + \dots + (2k-1)^2 + [2(k+1)-1]^2 = \frac{k(2k+1)(2k-1)}{3} + [2(k+1)-1]^2$$

$$= \frac{k(2k+1)(2k-1)}{3} + (2k+1)^2$$

$$= \frac{k(2k+1)(2k-1) + 3(2k+1)^2}{3}$$

$$= \frac{(2k+1)[k(2k-1) + 3(2k+1)]}{3}$$

$$= \frac{(2k+1)(2k^2+5k+3)}{3}$$

$$= \frac{(2k+1)(k+1)(2k+3)}{3}$$

$$= \frac{(k+1)[2(k+1)+1][2(k+1)-1]}{3}$$

and the statement is proved.

5. Prove $5 + 10 + \dots + 5n = \dfrac{5n(n+1)}{2}$

Step 1. Evaluating the formula when $n = 1$ yields

$$\frac{5 \cdot 1(1+1)}{2} = 5$$

as it should to verify the formula.

Step 2. For $n = k$ we have

$$5 + 10 + \dots + 5k = \frac{5k(k+1)}{2}$$

Adding the next multiple of 5 gives

$$5 + 10 + \dots + 5k + 5(k+1) = \frac{5k(k+1)}{2} + 5(k+1)$$

$$= \frac{5k(k+1)}{2} + \frac{10(k+1)}{2}$$

$$= \frac{5(k+1)(k+2)}{2}$$

$$= \frac{5(k+1)[(k+1)+1]}{2}$$

and the statement is proved.

8. Prove that $1^3 + 2^3 + 3^3 + \ldots + n^3 = \frac{n^2(n + 1)^2}{4}$

Step 1. Evaluating at $n = 1$ yields

$$\frac{1^2(1 + 1)^2}{4} = 1$$

as it should to verify the formula, since $1^3 = 1$.

Step 2. For $n = k$ we have

$$1^3 + 2^3 + 3^3 + \ldots + k^3 = \frac{k^2(k + 1)^2}{4}$$

adding the next consecutive cube yields

$$\begin{aligned} 1^3 + 2^3 + \ldots + k^3 + (k + 1)^3 &= \frac{k^2(k + 1)^2}{4} + (k + 1)^3 \\ &= \frac{k^2(k + 1)^2 + 4(k + 1)^3}{4} \\ &= \frac{(k + 1)^2(k^2 + 4k + 4)}{4} \\ &= \frac{(k + 1)^2(k + 2)^2}{4} \end{aligned}$$

and the statement is proved.

11. Prove that $a_n = a_1 + (n - 1)d$ for an arithmetic progression.

Step 1. Evaluating when $n = 1$ yields $a_1 + (1 - 1)d = a_1$ or the first term of the progression, as it should.

Step 2. For $n = k$ we have $a_k = a_1 + (k - 1)d$. We know that adding d to a_k will produce a_{k+1} for an arithmetic progression, so we add d to both sides:

$$\begin{aligned} a_k + d &= a_1 + (k - 1)d + d \\ &= a_1 + kd \\ a_{k+1} &= a_1 + [(k + 1) - 1]d \end{aligned}$$

as required to prove the statement.

14. Prove that $a + ar + ar^2 + \ldots + ar^{n-1} = \frac{a(1 - r^n)}{1 - r}$.

Step 1. Evaluating when $n = 1$ yields

$$\frac{a(1 - r)}{1 - r} = a$$

as it should to verify the statement.

(continued on page 222)

14. continued

Step 2. For $n = k$ we have

$$a + ar + \ldots + ar^{k-1} = \frac{a(1 - r^k)}{1 - r}$$

Adding the next term to both sides yields

$$a + ar + \ldots + ar^{k-1} + ar^k = \frac{a(1 - r^k)}{1 - r} + ar^k$$

$$= \frac{a(1 - r^k) + ar^k(1 - r)}{1 - r}$$

$$= \frac{a(1 - r^k + r^k - r^{k+1})}{1 - r}$$

$$= \frac{a(1 - r^{k+1})}{1 - r}$$

and the statement is proved.

Section 9.4 The Binomial Theorem

BASIC CONCEPTS you should learn from this unit.

1. The expansion of $(a + b)^n$ where n is any natural number is simply the result of carrying out the multiplication.
2. Pascal's triangle is one way of determining the coefficients of the terms of the expansion.
3. Another way to find the coefficients is the following: In the preceding term, multiply the coefficient by the exponent of a and then divide by one more than the exponent of b. This works, of course, for all terms except the first.
4. Employing the binomial theorem is another way, although this is essentially the same as concept 3.
5. The sum of the exponents in any term of the binomial expansion equals n.
6. The product of the first n natural numbers is n factorial, which in symbols is written $n!$.
7. $0! = 1$ by definition.

WHAT YOU SHOULD KNOW HOW TO DO upon completion of this unit.

1. Definine expansion of a binomial.
2. Define $n!$ (n factorial).
3. Expand a binomial using the binomial theorem.
4. Find a specific term in a binomial expansion.

SOLUTIONS to Exercise Set 9.4

2. $$(2a - 3b)^6 = (2a)^6 - 6(2a)^5(3b) + \frac{30(2a)^4(3b)^2}{2!} - \frac{120(2a)^3(3b)^3}{3!} + \frac{360(2a)^2(3b)^4}{4!} - \frac{720(2a)(3b)^5}{5!} + (3b)^6$$
$$= 64a^6 - 576a^5b + 2160a^4b^2 - 4320a^3b^3 + 4860a^2b^4 - 2916ab^5 + 729b^6$$

5. $$(2 - xy)^5 = 2^5 - 5\cdot 2^4xy + \frac{20\cdot 2^3(xy)^2}{2!} - \frac{60\cdot 2^2(xy)^3}{3!} + \frac{120\cdot 2(xy)^4}{4!} - \frac{120(xy)^5}{5!}$$
$$= 32 - 80xy + 80x^2y^2 - 40x^3y^3 + 10x^4y^4 - x^5y^5$$

8. $$(x - y)^7 = x^7 - 7x^6y + \frac{42x^5y^2}{2!} - \frac{5\cdot 42x^4y^3}{3!} + \frac{4(210)x^3y^4}{4!} - \frac{3(840)x^2y^5}{5!} + \frac{2\cdot(2520)xy^6}{6!} - \frac{5040}{7!}y^7$$
$$= x^7 - 7x^6y + 21x^5y^2 - 35x^4y^3 + 35x^3y^4 - 21x^2y^5 + 7xy^6 - y^7$$

11. $$\left(\frac{1}{3}x + 2\right)^3 = \left(\frac{x}{3}\right)^3 + 3\left(\frac{x}{2}\right)^2\cdot 2 + \frac{6(x/3)\cdot 2^2}{2!} + 2^3$$
$$= \frac{x^3}{27} + \frac{2x^2}{3} + 4x + 8$$

14. The first four terms of $(x - 3)^{12}$ are
$$x^{12} - 12x^{11}\cdot 3 + \frac{11\cdot 12x^{10}\cdot 3^2}{2!} - \frac{10\cdot 132x^9\cdot 3^3}{3!} = x^{12} - 36x^{11} + 594x^{10} - 5940x^9$$

17. The first four terms of $(2x - 3y)^{14}$ are
$$(2x)^{14} - 14(2x)^{13}(3y) + \frac{13\cdot 14(2x)^{12}(3y)^2}{2!} - \frac{12\cdot 13\cdot 14(2x)^{11}(3y)^3}{3!}$$
$$= 16{,}384x - 344{,}064x^{13}y + 3{,}354{,}624x^{12}y^2 - 20{,}127{,}744x^{11}y^3$$

20. The first four terms of $\left(x - \frac{1}{y}\right)^{15}$ are
$$x^{15} - 15x^{14}\cdot\frac{1}{y} + \frac{14\cdot 15x^{13}}{2!}\cdot\frac{1}{y^2} - \frac{13\cdot 14\cdot 15x^{12}}{3!}\cdot\frac{1}{y^3} = x^{15} - \frac{15x^{14}}{y} + \frac{105x^{13}}{y^2} - \frac{455x^{12}}{y^3}$$

23. $$\frac{12!}{11!} = \frac{12\cdot 11\cdot 10\cdot\ \ldots\ \cdot 2\cdot 1}{11\cdot 10\cdot\ \ldots\ \cdot 2\cdot 1} = 12$$

26. $$\frac{7!}{9!} = \frac{7\cdot 6\cdot 5\cdot\ \ldots\ \cdot 2\cdot 1}{9\cdot 8\cdot 7\cdot 6\cdot\ \ldots\ \cdot 2\cdot 1} = \frac{1}{72}$$

29. $\frac{6!}{3!} = 6\cdot5\cdot4 = 120$

32. $\frac{(n+1)!}{(n-1)!} = \frac{(n+1)(n)(n-1)\cdot\ ...\ \cdot3\cdot2\cdot1}{(n-1)(n-2)\cdot\ ...\ \cdot3\cdot2\cdot1} = (n+1)(n) = n^2 + n$

35. The fifth term of $(x/2 - y)^{12}$ is the term with y^4 and $(x/2)^8$ and the coefficient of $\frac{12!}{4!8!}$. Finally, then, we have $\frac{495}{256}x^8y^4$.

38. The next to last term of $(a + 4b)^5$ contains a and $(4b)^4$ and a coefficient of $\frac{5!}{4!1!}$ or $5a(256b^4) = 1280ab^4$.

41. We want the term containing $(3x)^4$ and $(4y)^3$ and a coefficient of $\frac{7!}{3!4!}$. Our term, then, is $35(81x^4)(64y^3) = 181{,}440x^4y^3$.

44. Since $(x^2)^4 = x^8$ we seek the term containing $(x^2)^4$ and $\left(\frac{1}{y}\right)^4$ with a coefficient of $\frac{8!}{4!4!}$ or $\frac{70x^8}{y^4}$.

47. $(1 + 0.3)^6 = 1^6 + 6\cdot1^5(0.3)\cdot + \frac{30\cdot1^4(0.3)^2}{2!} + \frac{120\cdot1^3(0.3)^3}{3!} + \frac{360\cdot1^2(0.3)^4}{4!}$

$$+ \frac{720\cdot1\cdot(0.3)^5}{5!} + \frac{720\cdot(0.3)^6}{6!}$$

$= 1 + 1.8 + 1.35 + 0.54 + 0.1215 + 0.01458 + 0.000729$

$= 4.8268$ rounded to four decimal places.

Section 9.5 Counting: Permutations and Combinations

BASIC CONCEPTS you should learn from this unit.

1. The fundamental counting principle is: If one event can occur in m ways and a second event can happen in n ways, then both events can happen in $m\cdot n$ ways.

2. A permutation is an arrangement that can be made using all or some of the elements of a set of objects without repetition.

3. The number of permuations of n objects taken r at a time is

$$P(n, r) = \frac{n!}{(n-r)!}$$

4. If we are interested in the number of ways of selecting objects and the order in which the objects are selected is not important, we use the word combinations to describe the situation.

5. The number of <u>combinations</u> of n objects taken r at a time is

$$C(n, r) = \frac{n!}{r!(n-r)!}$$

WHAT YOU SHOULD KNOW HOW TO DO upon completion of this unit.

1. Determine whether a problem is a permutation problem or a combination problem, or whether it involves both principles.

2. Calculate the number of permutations of n objects taken r at a time.

3. Calculate the number of combinations of n objects taken r at a time.

SOLUTIONS to Exercise Set 9.5

2. There are five letters in <u>study</u>, so the answer is 5! or 120 .

5. Each time the coin is tossed it has two possible ways to come up, and since there are eight events, or tosses, the answer is $2^8 = 256$.

8. $P(6, 6) = \frac{6!}{(6-6)!} = \frac{6!}{0!} = 6! = 720$.

11. $P(8, 3) = \frac{8!}{(8-3)!} = \frac{8!}{5!} = 8 \cdot 7 \cdot 6 = 336$.

14. $P(8, 4) = \frac{8!}{4!} = 8 \cdot 7 \cdot 6 \cdot 5 = 1680$.

17. $P(3, 1) = \frac{3!}{2!} = 3$.

20. a. Since there are ten people and no restrictions are imposed, we have simply $P(10, 10) = 10! = 3{,}628{,}800$.

b. The men and women would have to occupy alternate seats according to the condition; that is, M W M W M W M W M W or the reverse of this. We thus have $2 \cdot 5 \cdot 5 \cdot 4 \cdot 4 \cdot 3 \cdot 3 \cdot 2 \cdot 2 \cdot 1 \cdot 1$ as the solution, or we could say $2 \cdot 5! \cdot 5! = 28{,}800$.

23. Since the e's and the d's are not distinguishable, we would have

$$\frac{6!}{3!2!} = 60$$

26. Using the letters of MISSISSIPPI four at a time:

There are two possibilities involving four of the same letter; each can occur in one way for a total of two ways.

There are eight possibilities involving three of one letter and one of another letter; each can occur in $\frac{4!}{3!} = 4$ ways, for a total of 32 ways.

There are three possibilities involving two of one letter and two of another letter; each can occur in $\frac{4!}{2!2!} = 6$ ways, for a total of 18 ways .

26. continued

There are two possibilities involving one of one letter, one of another, and two of a third letter; each can occur in $\frac{4!}{2!} = 12$ ways, for a total of 24 ways.

Adding these, we have $2 + 32 + 18 + 24 = \underline{76}$ ways

29. $C(9, 3) = \frac{9!}{3!6!} = \frac{9 \cdot 8 \cdot 7}{3 \cdot 2} = 84$.

32. $C(7, 1) = \frac{7!}{1!6!} = 7$.

35. $C(n, n-1) = \frac{n!}{(n-1)! \cdot [n-(n-1)]!} = n$.

38. The faculty members can be chosen in $C(8, 2)$ or 28 ways, and the students in $C(10, 3)$ or 120 ways; $28 \cdot 120 = 3360$.

41. a. $26 \cdot 25 \cdot 24 = 15{,}600$.

b. $26 \cdot 26 \cdot 26 = 17{,}576$.

44. The basic counting principle is used here, yielding $3 \cdot 5 \cdot 2 = 30$ cars.

47. The four prisoners can be selected in $C(12, 4)$ or 495 ways, and the three officers in $C(10, 3)$ or 120 ways. The final answer, then, is $495 \cdot 120 = 59{,}400$.

50. a. Seven are defective and we can select four of these seven in $C(7, 4) = 35$ ways.

b. Two of the seven defectives can be selected in $C(7, 2) = 21$ ways; two of the five good transistors can be selected in $C(5, 2) = 10$ ways. Finally, there are $10 \cdot 21 = 210$ ways to select two good and two defective.

c. Five are good, so we have $C(5, 4) = 5$ ways to select four of these five.

d. $C(7, 3) \cdot C(5, 1) = 35 \cdot 5 = 175$ ways.

Section 9.6 Probability

BASIC CONCEPTS you should learn from this unit.

1. The probability of an event occurring is defined as

$$\frac{\text{number of successful outcomes}}{\text{total number of outcomes}}$$

2. A probability of 1 means that the event is certain to occur, and a probability of 0 means that the event cannot occur.

3. If p is the probability that an event will occur, then $1 - p$ is the probability that the event will not occur.

4. If two independent events have probabilities of p_1 and p_2, then the probability that both events will occur is $p_1 p_2$.

WHAT YOU SHOULD KNOW HOW TO DO upon completion of this unit.

1. Define probability.

2. Find the probability that an event will not occur if the probability that it will occur is known.

3. Compute the probability of an event happening using the principles of combinations, permutations, and other methods of counting.

4. Compute the probability that two or more events will happen if the individual probabilities are known.

SOLUTIONS to Exercise Set 9.6

2. The probability of a 5 not showing up is $\frac{5}{6}$ for each die. For both dice not to show a 5 is then $\frac{5}{6}\cdot\frac{5}{6} = \frac{25}{36}$. Now the probability of at least one 5 showing up is $1 - \frac{25}{36}$ or $\frac{11}{36}$.

5. a. Similar to Problem 2: $1 - \frac{25}{36} = \frac{11}{36}$.

   b. There are five ways a sum of eight can occur, so the probability of a sum of eight is $\frac{5}{36}$.

   c. The probability that a 3 or a 4 will not appear means that a 1, 2, 5, or 6 will occur. So for each die the probability we seek is $\frac{2}{3}$ and for both dice $\frac{2}{3}\cdot\frac{2}{3} = \frac{4}{9}$.

8. a. The women can be selected in $C(18, 3) = 816$ ways. The total number of ways to select the committee is $C(30, 3) = 4060$ ways. Thus we have $\frac{816}{4060}$ or $\frac{204}{1015}$.

   b. This is equivalent to having an all-male committee, which is $\frac{C(12, 3)}{C(30, 3)}$ by reasoning similar to that used for part (a). This is $\frac{220}{4060}$ or $\frac{11}{203}$

   c. The woman can be chosen in $C(18, 1) = 18$ ways and the two men in $C(12, 2) = 66$ ways; both events can occur in $18\cdot 66 = 1188$ ways. So the probability of occurrence is $\frac{1188}{4060} = \frac{297}{1015}$.

   d. This is equivalent to the committee not having all men; this is $1 - \frac{11}{203} = \frac{192}{203}$. (Refer to part (b) above.)

11. Since there are 13 hearts to start with, then 12, then 11, then 10, the probability we seek is

$$\frac{13}{52}\cdot\frac{12}{51}\cdot\frac{11}{50}\cdot\frac{10}{49} = \frac{17,160}{6,497,400} = \frac{11}{4165}$$

14. a. This would be $1 - 0.6 = 0.4$.

b. $(0.4)(0.4) = 0.16$.

17. That 2% are defective implies that 98% are good, and the probability of selecting one good one is 0.98.

a. $(0.98)^4 = 0.922$ approximately .

b. $(0.2)^4 = 1.6 \times 10^{-7}$.

SOLUTIONS to Review Exercises Chapter 9

2. Letting $n = 1, 2, 3$ in order, we have

$$\frac{1-1}{1+1}, \frac{8-1}{2+1}, \frac{27-1}{3+1} \quad \text{or} \quad 0, \frac{7}{3}, \frac{13}{2}.$$

5. $a_{21} = -2 + (20) \cdot 2 = 38$.

8. Since $a_{23} = a_1 + 22d$, we have $-15 = -4 + 22d$ and thus $d = -\frac{1}{2}$. Now $a_{26} = -4 + (25)(-\frac{1}{2}) = -\frac{33}{2}$.

11. For 2, -6, 18, -54 the common ratio is -3 .

14. Using $a_n = a_1 r^{n-1}$, we have $a_1 = \underline{-2}$; $a_2 = -2(-1) = \underline{2}$; $a_3 = -2(-1)^2 = \underline{-2}$; $a_4 = -2(-1)^3 = \underline{2}$.

17. $\frac{1}{72} = 3 \cdot r^3$ so that $r^3 = \frac{1}{216}$ and $r = \frac{1}{6}$; hence the two means required are

$$3 \cdot \frac{1}{6} = \frac{1}{2} \quad \text{and} \quad 3 \cdot \frac{1}{36} = \frac{1}{12}$$

Finally we have $3, \frac{1}{2}, \frac{1}{12}, \frac{1}{72}$.

20. $S = \frac{a}{1-r} = \frac{5}{1-\frac{1}{2}} = 10$.

23. $$(2x - y)^4 = (2x)^4 - 4(2x)^3(y) + \frac{12(2x)^2y^2}{2} - \frac{24(2x)y^3}{3!} + y^4$$
$$= 16x^4 - 32x^3y + 24x^2y^2 - 8xy^3 + y^4$$

26. $6! = 6 \cdot 5 \cdot 4 \cdot 3 \cdot 2 \cdot 1 = 720$

29. $P(6, 3) = \frac{6!}{3!} = 120$.

32. $C(n, n-1) = \frac{n!}{(n-1)!(n-n+1)!} = n$.

35. This is a combination problem since the order of selection is not important. Thus we have $C(10, 6) = \frac{10!}{6!4!} + \frac{10\cdot 9\cdot 8\cdot 7}{4\cdot 3\cdot 2\cdot 1} = 210$.

38. The two white pens can be selected in $C(4, 2) = 6$ ways and the total number of ways in which two pens can be selected from twelve is $C(7, 2) = 21$ ways. $\frac{6}{21} = \frac{2}{7}$ is the probability we seek.

41. There are six husband-wife teams and there are $C(6, 2)$, or 15, ways to choose two of these teams. Second, from twelve people there are $C(12, 4) = 495$ ways to choose four. So there is a probability of $\frac{15}{495} = \frac{1}{33}$ of selecting two husband-wife teams.

Chapter 9 - A SHORT QUIZ

1. An infinite sequence is a function whose domain is the set of all ___?___ .

2. A sequence that terminates is called ___?___ .

3. The fifth term of $a_n = \frac{2n}{n+1}$ is found by ___?___ .

4. A sequence whose successive terms differ from each other by a fixed number is called a(an) ___?___ sequence.

5. A recursive formula such as $a_n = a_{n-1} + d$ allows us to find any term in a sequence if we know the ___?___ term.

6. An arithmetic <u>series</u> is simply the ___?___ of an arithmetic sequence.

7. The number $0.33\overline{3}$ could be written as the infinite series $0.3 + 0.03 + 0.003 + \ldots$. In this case the series is ___?___ with a common ratio of ___?___ .

8. You already know that $0.33\overline{3} = \frac{1}{3}$. Verify this by finding the sum of the infinite series in problem 7 .

9. If we expand $(x + y)^4$, we get $x^4 + 4x^3y + 6x^2y^2 + 4xy^3 + y^4$. Find now the following <u>combinations</u>:

   $C(4, 1)$; $C(4, 2)$; $C(4, 3)$; $C(4, 4)$.

   What do you observe?

10. What do you suspect $C(4, 0)$ is? Test your guess by the combinations formula.

11. Solve for x : $x\cdot C(6, 4) = P(6, 4)$.

12. Selecting five committee members from a group of eight people is a combination problem because ___?___ .

13. In many states auto license plates consist of three letters followed by three numerals usually with repetitions allowed. By what principle would the number of possible plates be determined?

14. If your name had three letters, what is the probability that you would get a license plate with your name?

15. Phone numbers in this country consist of a three-digit area code, with the first digit non-zero, and then seven more digits. Repetitions are allowed. Theoretically, how many phone numbers are possible?

16. Probability is defined as the ___?___ divided by the ___?___ .

17. If the probability of an event occurring is x, what is the probability of it not occurring?

18. If a coin is tossed 25 times and comes up heads each time, what is the probability it will come up heads on the twenty-sixth toss?

19. What is the probability that a coin would come up heads 25 times in a row?

20. Empirical probability is probability obtained from ___?___ rather than from theory.

Quiz Answers Chapter 9

1. Natural numbers

2. Finite

3. Substituting 5 for n in $\frac{2n}{n + 1}$ and then simplifying

4. Arithmetic

5. Preceding

6. Sum

7. Geometric, 0.1

8. $S = \frac{0.3}{1 - 0.1} = \frac{0.3}{0.9} = \frac{1}{3}$.

9. $C(4, 1) = 4$; $C(4, 2) = 6$; $C(4, 3) = 4$; $C(4, 4) = 1$. These are the coefficients of the expansion $(x + y)^4$.

10. 1 $C(4, 0) = \frac{4!}{0!4!} = 1$.

11. $x = 2$

12. order of selection is not important.

13. counting principle

14. $\frac{1 \cdot 10^3}{26^3 10^3} = \frac{1}{26^3}$.

15. $9 \cdot 10^9 = 9{,}000{,}000{,}000$.

16. Number of successful outcomes, ... , total number of outcomes

17. $1 - x$

18. 1/2

19. $\left(\frac{1}{2}\right)^{25}$

20. Experience

Printed in France by Amazon
Brétigny-sur-Orge, FR

51786468R00100